Law, Human Agency and Autonomic Computing

Law, Human Agency and Autonomic Computing investigates the legal impli-cations of the notion and experience of human agency, implied by the emerging paradigm of autonomic computing and the socio-technical infrastructures it supports. The development of autonomic computing and ambient intelligence – self-governing systems – challenge traditional philosophical conceptions of human self-constitution and agency, with significant consequences for the theory and practice of constitutional self-government. Ideas of identity, subjectivity, agency, personhood, intentionality, and embodiment are all central to the functioning of modern legal systems. But once artificial entities become more autonomic, and less dependent on deliberate human intervention, notions like agency, intentionality and self-determination, become too fragile to serve as defining criteria for human subjectivity, personality or identity, and for charac-terizing the processes through which individual citizens become moral and legal subjects. Are autonomic – yet artificial – systems shrinking the distance between (acting) subjects and (acted upon) objects? How 'distinctively human' would agency be in a world of autonomic computing? Or, alternatively, does autonomic computing merely disclose that we were never, in this sense, 'human' anyway? A dialogue between philosophers of technology and philosophers of law, this book addresses these questions, as it takes up the unprecedented opportunity that autonomic computing and ambient intelligence offer for a reassessment of the most basic concepts of law.

Mireille Hildebrandt is a senior researcher at the Centre for Law, Science, Technology and Society Studies (LSTS) at Vrije Universiteit Brussel. She is Associate Professor of Jurisprudence at the Erasmus School of Law, Erasmus University Rotterdam and Full Professor of Smart Environments, Data Proection and the Rule of Law at the Institute of Computer and Information Sciences (ICIS) at Radboud University Nijmegen in the Nertherlands.

Antoinette Rouvroy is research associate of the National Fund for Scientific Research (FNRS) and senior researcher at the Information Technology and Law Research Centre (CRID) of the University of Namur, Belgium.

Law, Human Agency and Autonomic Computing

The philosophy of law meets the philosophy of technology

Edited by Mireille Hildebrandt and Antoinette Rouvroy

Routledge
Taylor & Francis Group

LONDON AND NEW YORK

First published 2011
by Routledge
2 Park Square, Milton Park, Abingdon, Oxon, OX14 4RN

Simultaneously published in the USA and Canada
by Routledge
711 Third Avenue, New York, NY 10017

Routledge is an imprint of the Taylor & Francis Group, an informa business
First issued in paperback 2013

British Library Cataloguing in Publication Data
A catalogue record for this book is available from the British Library

Library of Congress Cataloguing in Publication Data
    Law, human agency, and autonomic computing : the philosophy
    of law meets the philosophy of technology / edited by Mireille Hildebrandt
    and Antoinette Rouvroy.
        p. cm.
    "Simultaneously published in the USA and Canada."
    1. Technology and law. 2. Law--Philosophy. 3. Liberty. I. Hildebrandt, M.
    II. Rouvroy, Antoinette.
    K487.T4L39 2011
    344'.095—dc22   2010040194

ISBN13: 978–0–415–59323–6 (hbk)
ISBN13: 978–0–203–82834–2 (ebk)
ISBN13: 978–0–415–72015–1 (pbk)

Typeset in Times New Roman
by Keystroke, Station Road, Codsall, Wolverhampton

Contents

Acknowledgements

The thoughts and reasonings developed in this volume were tested during a small scale colloquium on 16 January 2009, which was a part of the larger Conference on Computing, Privacy and Data Protection 2009 (CPDP09) in Brussels.[1] All authors took part in this colloquium, as well as Rafael Capurro (philosopher of information ethics) who shared a paper entitled 'Towards a Comparative Theory of Agents' but preferred to publish this on his own website. As we live in a time of reconfiguration of the framework for the publication of scholarly undertakings, we respect his choice, though his text is now separated from the productive proximity of the other chapters. We wish to thank the organizers of the larger Conference, LSTS, CRID, TILT and INRA,[2] and express our special thanks to Paul de Hert, Serge Gutwirth, Rocco Bellanova and to Yves Poullet, for enabling the synergy of a reading panel that practises 'slow thinking' with the hectic succession of highly informative presentations on the intersection of law and computer science with regard to privacy and data protection. Grietje Gorus and Katrijn de Marez provided professional logical support for all participants. The reading panel was also part and parcel of the five-year focused research project (GOA) on 'Law and Autonomic Computing: Mutual Transformations' that was allotted to the Centre for Law, Science, Technology and Society studies (LSTS) at Vrije Universiteit Brussel, of which Mireille Hildebrandt is a co-author and co-ordinator, together with Serge Gutwirth (director of LSTS), Paul de Hert (initiator of the CPDP09), and Laurent de Sutter (researcher at LSTS). We also thank Katja de Vries and Niels van Dijk, PhD students within the GOA project on law and autonomic computing for their enthusiastic, critical as well as constructive interventions on the topic. Dymphy Schuurman, student-assistant at the Erasmus School of Law, provided help in editing the chapters. We admire the professional patience and stamina of Routledge's editorial assistant, Holly Davis, who helped us through the final editing process plus all that comes with it and, last but not least, we are immensely grateful to Colin Perrin for his trusting enthusiasm and support in getting this book published by Routledge.

Notes

1 See www.cpdpconferences.org.
2 The Centre for Law, Science, Technology and Society studies (LSTS) at Vrije Universiteit Brussel, Centre de recherche informatique et droit (CRID) of the Facultés universitaire 'Notre-Dame de la Paix' (FUNDP), Namur, Tilburg Institute for Law, Technology and Society (TILT) at Tilburg University and the National Institute for Research in Computer Science and Control (INRIA), Paris.

On the contributors

Roger Brownsword is Professor of Law and Director of TELOS, King's College London. He is also an Honorary Professor in Law at the University of Sheffield. In an academic career spanning 40 years, he has some 200 publications. His recent books include *Consent in the Law* (Hart, Oxford, 2007) (with Deryck Beyleveld), *Rights, Regulation and the Technological Revolution* (Oxford University Press, 2008) and *Regulating Technologies* (Hart, Oxford, 2008) (co-edited with Karen Yeung). He and Han Somsen act as general editors of the journal *Law, Innovation and Technology*, launched in 2009.

Jos de Mul studied philosophy, art history and law at the universities of Utrecht and Amsterdam. Since 1993 he has been full professor Philosophy of Man and Culture at the Faculty of Philosophy, Erasmus University Rotterdam, and scientific director of the Research Institute Philosophy of Information and Communication Technology (ICT). Among his book publications are: *Romantic Desire in (Post)Modern Art and Philosophy* (Albany, NY: State University of New York Press, 1999), *The Tragedy of Finitude. Dilthey's Hermeneutics of Life* (New Haven, Yale University Press, 2004), and *Cyberspace Odyssey. Towards a Virtual Ontology and Anthropology* (Cambridge Scholars Publishing, 2010).

Massimo Durante is Researcher in Philosophy of Law (Faculty of Law, University of Turin) and holds a PhD in Philosophy of Law (Faculty of Law, University of Turin) and a PhD in History of Philosophy (Faculty of Philosophy, University of Paris IV Sorbonne). His main fields of research are: Legal Informatics, Phenomenology and Ethics. He collaborates with the Institut d'Etudes Lévinassiennes (Paris-Jerusalem) and the reviews: *Cahiers d'Etudes Lévinassiennes*; *Rivista di Filosofia Politica*. Author of monographies and articles in different languages in the fields of Philosophy of Law, Phenomenology, Ethics and Legal Informatics, he has recently published the books: *Etica, Diritto, Decentramento. Dalla sussidiarietà digitale all'economia dell'informazione in rete* (Torino: Giappichelli, 2007); *Responsabilità di fronte alla storia. La filosofia di Emmanuel Lévinas tra alterità e terzietà* (Genova: Il Melangolo, 2008).

Mireille Hildebrandt is a senior researcher at the Centre for Law Science Technology and Society Studies (LSTS) at Vrije Universiteit Brussel, Associate Professor of Jurisprudence at Erasmus University Rotterdam, and Full Professor of Smart Environments, Data Proection and the Rule of Law at the Institute of Computer and Information Sciences (ICIS) at Radboud University Nijmegen in the Nertherlands. She is associate editor of *Criminal Law and Philosophy* and of *Identity in the Information Society (IDIS)* and from 2004 to 2009 coordinated research on profiling technologies within the FP6 Network of Excellence on the Future of Identity in the Information Society (FIDIS). In 2008 she published *Profiling the European Citizen. Cross-Disciplinary Perspectives*, co-edited with Serge Gutwirth and in 2010 she published 'The challenges of Ambient Law and Legal Protection in the Profiling Era' in *The Modern Law Review*, together with Bert-Jaap Koops.

Don Ihde is Distinguished Professor of Philosophy, Stony Brook University. He is also the Director of the Technoscience Research Group which hosts international Visiting Scholars, Postdocs and advanced PhD students. Its seminar focuses upon philosophies of science and technology and science studies and in recent years has produced a number of panels presenting 'postphenomenological research' projects. Don Ihde is the author of twenty books, including *Ironic Technics* (2008), *Listening and Voice* 2nd edition (2007), *Bodies in Technology* (2002), and *Postphenomenology: A Critical Companion to Ihde* was published in 2006, edited by Evan Selinger.

Jannis Kallinikos is Professor in the Information Systems and Innovation Group, Department of Management at London School of Economics. Major research interests involve the institutional construction of predictable worlds (that is, the practices, technologies and formal languages by which organizations are rendered predictable and manageable) and the investigation of the modes by which current institutional and technological developments challenge key forms and institutions that dominated modernity. Some of these themes are analyzed in detail in his recent book *The Consequences of Information: Institutional Implications of Technological Change*, Edward Elgar, 2006.

Hyo Yoon Kang is a postdoctoral research fellow at the Max Planck Institute for the History of Science in Berlin where her current project is concerned with the relationship between patent law and biological taxonomies. Her research interests combine property, social and legal theory, science studies, anthropology and intellectual property. She completed her PhD in Law at the European University Institute in Florence on a dissertation, entitled 'Processes of Individuation and Multiplicity: the Human Person in Patent Law Relating to Human Genetic Material and Information'. Previously, Hyo Kang was a teaching fellow in property law at the London School of Economics.

Paul Mathias is currently a Professor of Philosophy at the Lycée Henri IV in Paris. From 1990 to 2001, he was an assistant professor at the Institut d'Etudes

Politiques de Paris, and was nominated in 2004 for the position of Program Director at the Collège International de Philosophie (Paris). Among other works related to the history of Philosophy and culminating with the recently published *Montaigne ou l'usage du Monde* (Vrin, 2006), Paul Mathias wrote *La Cité Internet* (Presses de Sciences-Po in 1997) and *Des Libertés numériques* (P.U.F., 2008). Since 1995 he has also participated in a research group at the École Normale Supérieure (Paris), 'Réseaux, savoirs, et territories', for which he has organized colloquia on Internet uses (1999), Measurements of the Internet (2003), and Traditional vs. computer writing (2008). Since 2005, he has also been a member of the scientific committee of "Vox Internet", a working group at the Maison des Sciences de l'Homme (Paris).

Stefano Rodotà is Professor of Law, University of Roma 'La Sapienza'. Chair, Scientific Committee of the Agency for Fundamental Rights, European Union. Chair, Internet Governance Forum Italy. Former President of the Italian Data Protection Commission and of the European Group on Data Protection. He is a Member of the Convention for the Charter of Fundamental Rights of the European Union, Visiting Fellow, All Souls College, Oxford. Visiting Professor, Stanford School of Law. Professeur à la Faculté de Droit, Paris 1, Panthéon-Sorbonne. Laurea honoris causa Université « Michel de Montaigne », Bordeaux. Former Member of the Italian and European Parliament, of the Parliamentary Assembly of the Council of Europe. Among his books: *Tecnologie e diritti* (Bologna, 1995); *Tecnopolitica* (Roma-Bari, 2004, translated into French and Spanish); *Intervista su privacy e libertà* (Roma-Bari, 2005); *La vita e le regole. Tra diritto e non diritto* (Milano, 2006); *Dal soggetto alla persona* (Napoli, 2007).

Antoinette Rouvroy is FNRS (National Fund for Scientific Research) research associate and researcher at the Information Technology and Law Research Centre (CRID) of the University of Namur, Belgium. She is particularly interested in the mechanisms of mutual production between sciences and technologies and cultural, political, economic and legal frameworks. She is the author of Human Genes and Neoliberal Governance: A Foucauldian Critique, Abingdon and New-York: Routledge-Cavendish, 2008, and of numerous articles about the ethical, legal and political challenges raised by the new information, communication and surveillance technologies (biometrics, RFIDs, ubiquitous computing, ambient intelligence, persuasive technologies, . . .) and their convergence. Most recently, her publications have been focused on statistical and algorithmic governmentality.

Bibi van den Berg studied Philosophy at Erasmus University Rotterdam and defended her dissertation entitled 'The Situated Self: Identity in a World of Ambient Intelligence' at the same university in 2009. In her dissertation she researched the consequences of the realization of the technological vision of the near future of Philips and the European Commission, entitled Ambient

Intelligence, for the construction and expression of identities. Presently she is a postdoc in philosophy of technology at the Tilburg Institute for Law, Technology and Society (TILT) of Tilburg University. She researches: (1) social/legal issues surrounding autonomous technologies, and (2) identity and privacy in online worlds.

Peter-Paul Verbeek (1970) is associate professor of philosophy at the Department of Philosophy of the University of Twente, and director of the international master program Philosophy of Science, Technology and Society. His research focuses on the social and cultural roles of technology and the ethical and anthropological aspects of human-technology relations. Currently he is working on a project about the ethical and anthropological aspects of biotechnology (NWO VIDI grant 2007), having just finished a project about the moral significance of technologies, and its implications for ethical theory and the ethics of design (NWO VENI grant 2003). He recently published the book What Things Do: Philosophical Reflections on Technology, Agency, and Design (Penn State University Press, 2005).

Foreword

> The situation was rather analogous to what might be described as sleep in human beings, but there were no dreams. The awareness of George Ten and George Nine was limited, slow, and spasmodic, but what there was of it was of the real world.
>
> They could talk to each other occasionally in barely heard whispers, a word or syllable now, another at another time, whenever the random positronic surges briefly intensified above the necessary threshold. To each it seemed a connected conversation carried on in a glimmering passage of time.
>
> "Why are we so?" whispered George Nine.
>
> "The human beings will not accept us otherwise," whispered George Ten, "They will someday."
>
> "When?"
>
> "In some years. The exact time does not matter. Man does not exist alone but is part of an enormously complex pattern of life forms. When enough of that pattern is roboticized, then we will be accepted."
>
> Isaac Asimov, '. . .That Though Art Mindful of Him'

The short story that includes the above passage is required reading in a course that I teach at the University of Ottawa called, *The Laws of Robotics.*

The plot is quintessential Asimov. Decision-makers at US Robots and Mechanical Men Inc. are seeking to change the laws prohibiting the use of robots on earth (robots are allowed only on space stations and extra-planetary mining operations). Changing the legal order, they realize, is predicated on a major cultural shift since most humans do not trust the robots. While the machines are hard-wired to obey human beings and not to harm them, delicate situations may nonetheless arise in the event of conflict *between* human beings. To navigate sticky situations, the robots require decision-making protocols to be coded into their positronic brains, determining which human beings are to be obeyed or unharmed. Of course, this requires robots to be capable of determining, all other things equal, which of the humans are *more equal* than the others. Resolving the complexities involved in such decision-making is a monumental task – one that, Harriman, Director of Research, secretly delegates to a robot named 'George Ten' and the companion model from which he was upgraded, 'George Nine'.

Prior to our investigation of the prudence in placing the entire fate of humanity into the hands of autonomic machinery (or the general strategy of creating enormously complex machines in order to have them solve enormously complex problems generated by the propagation of still other complex machines), my students and I start out by trying to unpack the significance of the passage cited at the outset, which I have always loved for its existential tone.

This dialogue between the two Georges (which gets even better as they amble, autonomically, down roads lesser travelled) makes poor Hamlet's 'To sleep, perchance to dream' look, well, robotic by comparison. Readers are inspired to imagine the grander implications of an increasing pattern of automation. Yet, as the two sleeping Georges reveal, the issue of robot consciousness is really a red herring. What ultimately matters is the moral question about human agency, about what humans are willing to permit and what humans are willing to accept.

In case you don't know how the story ends – *spoiler alert* – the two Georges delve even deeper, beyond their own existence, to discuss what it means to be human. In fact, this is part of their mandate; they are required to interpret the second law of robotics, namely that 'a robot must obey the orders given it by a human being. . .' In the end, adding insult to irony, Georges Nine and Ten determine that they too are human beings in all ways that truly matter. Given their superior reasoning and problem-solving abilities, they conclude that any commands issued by beings such as themselves would in fact take priority over the orders of biological humans.

Raised on a steady diet of *The Terminator, The Matrix, X-Men* and the like, when my law students read Asimov, they get geared up about the ethical and legal implications of strong artificial intelligence, the possibility of robot consciousness and the subsequent inevitability of robot uprisings.

Instead, I encourage them to start with more modest questions such as: what is the significance of Asimov's title? Usually a handful of students will have discovered for themselves that it is a reference to *Psalm 8:4*: 'What is man that thou art mindful of him?' An exceptional student once noted that the missing part of the Psalm in Asimov's title – 'What is Man?' – is also the name of a work by Mark Twain wherein an old man and his younger interlocutor engage in a parallel albeit more rigorous philosophical discussion of the sort had by Georges Nine and Ten. But, by the time we are done with the story, my students *all* come to realize just how much can be learned about questions of human agency – about humanity – through a philosophical assessment of the human project called 'autonomic computing'.

The holders of this esteemed volume, *The Philosophy of Law meets the Philosophy of Technology: Autonomic Computing and Transformations of Human Agency,* will soon discover that they need no longer rely on Asimov's science fiction to guide their thinking about these vexing philosophical issues. No longer limited to flights of fancy or thought experiments, readers will realize that vast industries have already dedicated themselves to the vision of building not android robots but 'self-aware' computing systems as the only viable solution to the lack of skilled humans otherwise tasked with managing the network of networks. Like their namesake – the autonomic nervous system – these sophisticated dualisms of hardware and software will carry out crucial regulatory and management decisions without any conscious recognition or effort from the unmoved prime-moving human beings who set them into motion, let alone from those who don't even realize that autonomic computing exists, nor that this is the means by which important determinations about their life-chances and opportunities are being made. Turning Asimov's phrase, those who read this insightful volume, just like those who wrote it, will confront the following question:

What are these machines that we are *unmindful* of them?

And, as with Asimov, reflections about the machines turn out to be reflections about us.

Perhaps in the spirit of the autonomic computing paradigm itself, readers may be completely unaware of the 'subliminal interventions' by the outstanding, coruscating editors of this volume, Mireille Hildebrandt and Antoinette Rouvroy. As someone who was invited but unable to participate in this project, I happen to know (even before I read it) that this book is *not* the typical motley crew of unrelated essays haphazardly assembled in response to a massive academic call for papers. The book results from a series of carefully orchestrated meetings, subsequent communications and much hard work; it reflects the commencement of an interactive dialogue by people who have a multifaceted command of the philosophies underlying the technologies and the regulatory oversight mechanisms at play.

In your hands is an artfully integrated, profound and meticulous collection of thinking about how autonomic computing affects traditional notions of agency and whether it will enhance or diminish legal accountability or our capacity for deliberate intentional action. This authoritative yet recalcitrant collection, featuring a cohesive, handpicked team of philosophers of law and technology, provides a choir of unique voices. The philosophers of technology set the stage through a series of provocative and enlightening investigations of the mutual constitutions of humans and their technological environment; the legal philosophers offer sophisticated consideration of the kind of responses and reconfigurations that are legally warranted.

Autonomic computing aside, one of the central goals of the project has been to spark a reciprocal and mutually beneficial interaction between philosophers of technology and the philosophers of law. As someone standing at the crossroads between these two disciplines, I am grateful to Mireille Hildebrandt and Antoinette Rouvroy for doing so. Those who read this book will know, as I do, that they have been extremely successful in carrying this out. In this regard, and in others, their collection is sure to make a lasting contribution, adding new layers of academic rigour and reflection to the existential queries initiated long ago by Asimov's sleeping robots.

Ian Kerr
Canada Research Chair in Ethics, Law and Technology
University of Ottawa

Introduction

A multifocal view of human agency in the era of autonomic computing

Mireille Hildebrandt

The objectives of this volume are threefold. First, it aims to investigate how different strands of the philosophy of technology can provide novel insights for the philosophy of law in an era of rapid technological change. Second, it seeks to explore how different approaches to philosophy of law can help philosophy of technology to come to terms with the normative impacts of technological devices and infrastructures. Third, we hope to provide a variety of perspectives on how autonomic computing impacts human agency, looking into the hybridization of humans and non-human artefacts and the sustainability of the notion of the liberal subject that informs mainstream philosophy of law.

Philosophy of law and of technology

This project was initiated with the idea that the current technological changes are not neutral with regard to the *instrumental* and *protective* dimensions of the law. For instance, the ease of peer to peer sharing technologies has decreased copyright law's ability to create an artificial scarcity, which is one of the *instrumental* dimensions of copyright law, whereas the rigidity of digital rights management has decreased copyright law's fair use exceptions (Lessig 2006), which are an example of law's *protective* function. However, as philosophers of technology have argued, though technology is never neutral, it is neither good nor bad (Kranzberg 1986). This position should enable us to steer clear of both the utopian visions that have plagued research into Artificial Intelligence (currently Kurzweil 2005) and the dystopian visions of Technology of some of the leading continental philosophers of technology (notably Heidegger 1977; Ellul 1967).[1]

In fact, an entire strand of American philosophy of technology nourishes on this nuanced position, articulating an empirical turn that aims to investigate the actual implications of specific technological devices and infrastructure (Achterhuis 2001).[2] This empirical turn builds on American pragmatism (Hickman 1990) and a variety of continental philosophical traditions, involving phenomenological, critical, hermeneutical and poststructuralist concerns. Concurrently, Science Technology and Society (STS) studies have developed their own strand of empirical inquiry into the myriad entanglements between people and the complex artefacts they construct and engage with (Rip et al. 1995).

According to Ihde (1990) the use of material tools (or technologies) is typical for human beings, just like the use of language.[3] Though law is also a tool, it is not a technology in this sense. In its primary dependence on language it seems to differ from technological tools, which have a materiality that co-determines their affordances.[4] Oral legal traditions share the ephemeral qualities of spoken language. However, since law has been embodied in technologies like those of the written and printed script, it shares the affordances of the script (eg Collins and Skover 1992; Hildebrandt 2008a; Vismann and Winthrop-Young 2008; Hildebrandt and Koops 2010). The characteristics of a written law differ substantially from those of an oral law (Glenn 2004), and many authors suggest that the transition to the digital era will again provoke far-reaching transitions to the law. We believe that by listening to the findings of philosophers of technology, legal philosophers may develop a novel sensitivity to the constitutive impact of law's current and future technological articulation.

This raises the issue of the implications of increasingly smart technologies for the current foundations of the legal system. Our main concern here is how the development of autonomic computing environments would affect the kind of human agency that is presumed by our legal order. Philosophers of technology have done some interesting research into how human agency is shaped by the technologies it uses. This type of research bypasses standard notions of a decontextualized rational subject, claiming that 'we have never been modern' (Latour 1993), that we have always been cyborgs or posthumans (Haraway 1991; Hayles 1999) and that our mind is not contained within our brains since it embraces technical devices as cognitive resources (Clark 2003). This invites us to rethink the idea of the individual human subject born with a free will and capable of deliberate intentional action, which is often thought to be the hallmark of liberal democracy and the basic assumption of Western legal systems. It makes sense, therefore, to investigate what philosophers of technology have to say about the mutual constitution of humans and their technological environments.

The normative implications of technological innovation have not always been on the forefront of research in the philosophy of technology (Rip 2003). This may be due to a fear of being associated with either utopian or dystopian visions of a reified Technology. Reiterating the idea that though 'technology is neither good nor bad, it is never neutral', we contend that there is an urgent need to assess the normativities triggered by technological change without however falling prey to moralism. Tracing potential normative impacts means investigating what types of behaviours are invited or inhibited, enforced or ruled out by a particular technological device or infrastructure (Hildebrandt 2008b; Verbeek 2005, 2006). This line of research is clearly related to research into the mediation of perception and cognition that is performed by the technologies we use and live with. The moral evaluation of these normative impacts is another matter, and though it requires a delineation of how behavioural patterns are reconfigured by a specific technology

we should not leap into moral condemnation or celebration before carefully investigating the normative impacts of specific technologies. A philosophy of technology
that is aware of the normative implications of specific technological innovations
could benefit from the practical demands that inform legal research, because – other
than ethics and moral philosophy – law forces one to take a position and consider
the practical consequences. For precisely this reason legal philosopher Solum has
contributed to the discourse on whether artificial intelligences are 'really' intelligent
by investigating whether they could function as a trustee and whether they might
qualify for constitutional protection (Solum 1992: 1232–33):

> First, putting the AI debate in a concrete legal context acts like an Occam's
> razor. By reexamining positions taken in cognitive science or the philosophy
> of artificial intelligence as legal arguments, we are forced to see them anew
> in a relentlessly pragmatic context.
>   Second, and more controversially, we can see the legal system as a repos-
> itory of knowledge, a formal accumulation of practical judgements. The law
> embodies core insights about the way the world works and how we evaluate
> it. (. . .) Hence, transforming the abstract debate over the possibility of AI into
> an imagined hard case forces us to check our intuitions and arguments against
> the assumptions that underlie social decisions made in many other contexts.

This volume provides such a hard case: how does autonomic computing affect
traditional notions of agency; will autonomic computing diminish or increase
individual legal accountability for harm caused; will it decrease or enhance our
capacity for deliberate intentional action? In the next section we will briefly discuss
the background of the third objective of this volume, by exploring the meaning of
the concepts of autonomic computing and human agency.

Autonomic computing and human agency

Many of the authors of this volume will explore the notions of autonomic computing and human agency within their own chapter and we like to emphasize that
our aim is not to suggest that there is consensus about the technical meaning of
autonomic computing or on the conceptual reach of the notion of human agency.
A plurality of meanings, however, does not imply that anything goes. We present
these notions as terms that denote a set of phenomena manifesting a family resem-
blance, rather than thinking in terms of general concepts that denote phenomena
necessarily sharing a common denominator.

Autonomic computing

Autonomic computing has been launched in 2001 by IBM as a vision on the new
computing paradigm. To explain what is meant IBM uses the metaphor of the
autonomic nervous system (Horn 2001):

The Autonomic Nervous System at Work

Think for a moment about one such system at work in our bodies, one so seamlessly embedded we barely notice it: the autonomic nervous system. It tells your heart how fast to beat, checks your blood's sugar and oxygen levels, and controls your pupils so the right amount of light reaches your eyes as you read these words. It monitors your temperature and adjusts your blood flow and skin functions to keep it at 98.6°F. It controls the digestion of your food and your reaction to stress—it can even make your hair stand on end if you're sufficiently frightened. It carries out these functions across a wide range of external conditions, always maintaining a steady internal state called homeostasis while readying your body for the task at hand.

But most significantly, it does all this without any conscious recognition or effort on your part. This allows you to think about what you want to do, and not how you'll do it: you can make a mad dash for the train without having to calculate how much faster to breathe and pump your heart, or if you'll need that little dose of adrenaline to make it through the doors before they close.

It's as if the autonomic nervous system says to you: 'Don't think about it—no need to. I've got it all covered'. That's precisely how we need to build computing systems—an approach we propose as autonomic computing.

Building on this organic metaphor, IBM has formulated the following eight requirements, introducing a curious mix of computer science terminology and anthropomorphic speculation:

1   To be autonomic, a computing system needs to 'know itself' – and comprise components that also possess a system identity.
2   An autonomic computing system must configure and reconfigure itself under varying and unpredictable conditions.
3   An autonomic computing system never settles for the status quo – it always looks for ways to optimize its workings.
4   An autonomic computing system must perform something akin to healing – it must be able to recover from routine and extraordinary events that might cause some of its parts to malfunction.
5   A virtual world is no less dangerous than the physical one, so an autonomic computing system must be an expert in self-protection.
6   An autonomic computing system knows its environment and the context surrounding its activity, and acts accordingly.
7   An autonomic computing system cannot exist in a hermetic environment.
8   Perhaps most critical for the user, an autonomic computing system will anticipate the optimized resources needed while keeping its complexity hidden.

IBM explains the need for autonomic computing in terms of the complexity of computing systems' management: the proliferation of computing devices and the increase in their computing powers increases the lack of skilled professionals who can manage all these interacting computing systems. They argue for smart computing instead of a further extension of computing devices and power, claiming that self-management will be the only viable solution.

Though autonomic computing is focused on the internal workings of computing systems, there are many analogies with earlier visions like those of ubiquitous computing (Weiser 1991) and Ambient Intelligence (Aarts and Marzano 2003; Aarts and Grotenhuis 2009; ISTAG 2001). Focal points of interest are hidden complexity, pervasive computing, context awareness, real time monitoring, proactive computing, and smart human-machine-interfacing. The idea that our wired interconnected environment will proactively cater to our inferred wishes is part and parcel of the ITU's vision of the Internet of Things (ITU 2005) and forms the very core of Ambient Intelligent environments. What autonomic computing adds is the idea of the systems' self-awareness in the sense of being able to operate at various levels, with higher levels capable of monitoring the performance of lower levels and intervening to repair or reconfigure elements of the system to make the whole system ever more robust. *Self-awareness*, then, may be the novel paradigm for smart technologies.

Though scepticism is warranted with regard to the performance of these systems, its central paradigm calls for scrutiny irrespective of whether they actually provide what is promised. The reason for such scrutiny is that even if a smart environment does not adequately infer our preferences, we will never find out insofar as intelligent environments operate at a subliminal level. Even if they are based on inaccurate and incomplete data or on inadequate inferences, we may end up accepting the decisions taken by the smart infrastructure as a convenient shortcut in the labyrinth of options provided by our ICT infrastructures. It might thus be that *subliminal intervention* is the primary novel paradigm of autonomic computing, next to its claimed *self-awareness*.

Human agency

The subliminal operation of autonomic computing raises issues around our notion of human agency. Obviously, the concept of 'human agency' is a prime example of an essentially contested concept (Gallie 1956), meaning that there is no consent about the content of the notion between different users of the term. Analytically oriented philosophers may assume that agency is about a person's capacity for practical reason, based on her beliefs about the world and her first and second order desires (cf Davidson 2001; Dworkin 1991; Morse 2008). Agency, from that perspective, is a matter of degree, just like autonomy. Phenomenologically oriented philosophers may wish to emphasize the embodiment of agents and the constraints as well as the possibilities this entails (cf Bourgine and Varela 1992; Dreyfus 1992). Within philosophy of technology a postphenomenological position has been developed by Ihde, introducing the concept of mediation to highlight the extent to which human agency is co-constituted by the technologies that mediate her perception and cognition (Ihde 1990; Verbeek 2005). Ihde extends the phenomeno-logical framework with a hermeneutical approach, demonstrating the multistability of technological devices that allows for different 'readings' of reality and thus for different impacts on the agent's 'reading' of herself. With Latour (1993), Haraway (1991) and Hayles (1999) one could question the separation of human agency from technological devices, arguing for a hybrid understanding of agency that rejects the idea of a rational agent that stands apart from the world.

Poststructuralist orientations may discard the notion of human agency altogether, declaring the death of the subject, equating subjectivity with subjection instead of emancipation (a position often attributed to Foucault, but see Rouvroy and Verbeek in this volume) or deconstructing the subject as a product of textual interpellations (Derrida). In fact, however, many poststructuralist accounts are less straightforward in their rejection of human agency, offering alternative pathways to a more rela-tional and relative understanding of a subject's identity and agency. This includes a processual view of identity in the sense that we never 'are', but always 'become', often articulated with the notion of identity performance (cf Butler 2006).

Another interesting notion of human agency can be mined from Rodotà's text in this volume. Rodotà reclaims the humanistic ideal of human identity even though acknowledging that it faces a myriad of threats from the encroaching data mining machines that recombine and correlate decontextualized data. He warns that such iterative de- and re-contextualization destabilizes the idea that human agency celebrates the construction of one's own identity in close relation with other human agents.

Evidently there are many implicit conceptions that inform everyday discourse, policy-making and legal practice. These tacit notions seem to share with analytical orientations the idea that human agency is sensitive to practical reason in the most pragmatic sense of the phrase, implying that we can hold each other to account on the basis of an expectation of reasonableness. This of course begs the question of what is to count as reasonable and the question of who decides – in the end – what is reasonable. Autonomic computing may create a situation in which smart com-puting environments decide what is reasonable, without them being capable of explaining their behaviours in terms of reasons, since their decisions are based on calculations and correlations. The issue at stake in this volume is whether this is a problem, how it could affect the kind of agency the law presupposes, and what kind of responses and reconfigurations this warrants in the field of legal philosophy.

Outline of a novel dialogue

The exploratory endeavours of the contributors have allowed us to identify a series of questions and a variety of approaches. The following outline does not exhaust the complexity of the issues nor the rich variety of approaches gathered in the volume and is merely a foretaste of the emerging debates and findings inaugurated in this book. We have grouped the chapters loosely on the basis of their philo-sophical lineage to either empirical and analytical or poststructuralist inspirations, acknowledging interstitial negotiations, borrowing and cross-border activities. The first part is more or less indebted to more empirical and analytical traditions (including phenomenological and hermeneutical research); the second part is inspired by poststructuralist traditions (integrating Hayles' perspective on bodies and embodiment); and the third part simply celebrates border-crossing (eg between neuroscience, information systems theory and humanistic ideals of human agency).

Part I: human agency shaped by smart technologies?

In Chapter 1 'Smart? Amsterdam urinals and autonomic computing', Don Ihde observes that in the narratives on smart environments, proactive technologies are supposed to 'nudge' or channel human behaviours in a certain (constantly re-calculated) direction. He contends that this creates a *double hermeneutic*, since the technologies try to calculate how humans will and how they had better act, while humans try to figure out how the technologies figure them out. This raises the question of to what extent such machine profiling will in fact match the ingenuity of human agency disclosed in examples like the famous fly in Amsterdam urinals. And even if so, Ihde argues that we have good reason to be sceptical about whether this double hermeneutic will seriously transform human agency.

In Chapter 2 'Subject to technology', Peter-Paul Verbeek argues that being human implies technological mediation and requires rethinking *ethics beyond the autonomous moral agent*. He then investigates what kind of ethical framework could do justice to the technologically mediated character of human existence and to the *interwoven character of morality and technology*.

In Chapter 3 'Human autonomy in the age of computer-mediated agency' Jos de Mul and Bibi van den Berg contend that to a considerable extent, human action has always been 'remote controlled' by internal and external factors which are beyond individuals' control. They argue that it is the reflection on such remote control *a posteriori* that allows for a 'reflexive appropriation' of these factors as our own motivators. The question they thus raise is what difference autonomic computing makes at this point and under what circumstances it will either strengthen or hinder human agency, defined in terms of 'reflexive appropriation'.

In Chapter 4 'Autonomy, delegation, and responsibility', Roger Brownsword examines what type of autonomic computing environments (ACEs) enhance, erode or eliminate the characteristics that we associate with human *autonomy* and agency and how these ACEs will impact the possibility for agents to create and sustain communities of rights.

Part II: traces, embodiments and novel constitutions of the real

In Chapter 5 'Rethinking human identity in the age of autonomic computing: The philosophical idea of trace', Massimo Durante wonders to what extent and in which ways interacting with a proactive as well as responsive multi-agent environment impacts on individual autonomy and responsibility. His main point is that the locus of imputation in smart environments is the *trace* left by interacting agents, be they human or artificial, challenging traditional notions of agency such as the Cartesian subject, the Lockean person, the Husserlian transcendental ego, the Merleau-Pontean empirical body, or the Heideggerian Dasein.

In Chapter 6 'Autonomic computing, genomic data and human agency: The case for embodiment', Hyo Yoon Kang contends that the mutual implication of molecular biology, human genetics and autonomic computing generates a further erosion of the boundaries between the biological, corporeal and private as opposed to the social, informational and public. This leads her to explore how we could rethink legal personhood and human agency, taking into account N. Katherine Hayles' notion of *embodiment*.

In Chapter 7 'Technology, virtuality and utopia: Governmentality in an age of autonomic computing' Antoinette Rouvroy puts forward that autonomic computing reinforces an increasingly *statistical governance of the* 'real', exemplifying an epistemic change in our relation to the 'real'. She explores the manner in which such a novel perceptual regime may affect the 'virtuality' that is characteristic for human *agency*.

Part III: accountability, autonomy and identity in the era of autonomic computing

In Chapter 8 'Autonomic and autonomous "thinking": Preconditions for criminal accountability', Mireille Hildebrandt investigates how *autonomic thinking*, as seen by cognitive science, relates to autonomous human action, preparing the question of how autonomic computing could manipulate our autonomic behaviours to the extent of eroding our capacity for autonomous action.

In Chapter 9 'Technology and Accountability: Autonomic computing and human agency', Jannis Kallinikos discusses how the collection and processing of data far beyond the capacity of the human mind impact the human capacity of autonomous action, and how data processing could contribute to the project of *legible trajectories of action*, thus enhancing the human capacity for autonomous action.

In Chapter 10 'Of machines and men: The road to identity. Scenes for a discussion', Stefano Rodotà remarks that autonomic computing seems to increase the *invisible control over the construction of identities*. He takes focus on how we could protect human identity if individual identities are increasingly determined by sources external to the person concerned and discusses how the law does and should react to the event of *hybriditized identities*. He demonstrates how recent judicial 'scenes' attest to the challenges now facing legislators and courts in dealing with the unprecedented intrusions of technological artefacts in the process of physical and intellectual self-determination.

In Chapter 11 'The BPI nexus', Paul Mathias moves into the different manners in which autonomic computing impacts on traditional notions of the *body* as an entity distinct from the technologies it uses; of the *person* as a formal substance presumed as the unity that allows for the attribution of liability; and of *identity* as a discrete subject that is the centre of gravity for autonomous action.

Endings

An introduction should not take away the reader's appetite by revealing too much of what follows. This volume advocates 'slow thinking' on a topic that may soon – or does already – affect all of us. There is no short-cut to its content, which requires the reader to fall in step – if only while reading – with narratives and lines of argument from different philosophical genres. And even if one manages to fall in step for some time, we all know that recalcitrance will make its appearance, generating novel dialogues, if only in the head of the reader.[5] This is what the technology of the script and the printing press afford: a distantiation that allows for a change of mind, an iterative reconstruction of the self in the face of whatever one does or does not agree with. Autonomic computing may enlarge the scope for such distantiation and reconstruction or creep under our skin by providing us with whatever it calculates us to want, even before we have become aware of this.

Notes

1 Which is not meant to suggest that the writings of these authors are not pertinent for the subject.
2 Achterhuis (2001) discusses Albert Borgmann, Hubert Dreyfus, Andrew Feenberg, Donna Haraway, Don Ihde and Langdon Winner.
3 This does not imply that animals don't use tools, nor that animals have not developed some form of language. However, the manner in which humans make use of physical and mental tools seems pertinent to their specificity. Cf. Ambrose 2001.
4 Ihde (1993:47–48) differentiates between a technology (defined as having a material component) and a technique (defined as a style, mode, habit of doing things or a calculative or rational method). In German this understanding of 'technology' would translate as 'Technik', and to complicate matters further some authors understand 'technology' as the reflection on 'technique' (cp. methodology as the reflection on methods). Foucault's usage of *Technologies of the Self* (Hutton et al. 1988) refers to methods, techniques that are not necessarily material. In that sense law could also be a technology.
5 See the epilogue by Antoinette Rouvroy in this volume for an elaboration of human agency as recalcitrance.

References

Aarts, E. and Grotenhuis F. (2009) 'Ambient Intelligence 2.0: Towards Synergetic Prosperity', in *AmI 2009: Towards Synergetic Prosperity*, Tscheligi, M., De Ruyter, B., Markopouluset, P. Wichert, R., Mirlacher, T., Meschtscherjakov, A., Reitberger, W. (eds), Berlin Heidelberg: Springer: 1–13.

Aarts, E. and Marzano, S. (eds) (2003) *The New Everyday. Views on Ambient Intelligence*, Rotterdam: 010.

Achterhuis, H. (ed.) (2001) *American Philosophy of Technology. The Empirical Turn* (translated by Crease, R.P.), Bloomington and Indianapolis: Indiana University Press.

Ambrose, S. H. (2001) 'Palaeolithic Technology and Human Evolution', *Science* 291: 1748–53.

Bourgine, P. and Varela, F.J. (1992) 'Towards a Practice of Autonomous Systems', in *Towards a Practice of Autonomous Systems. Proceedings of the First European Conference on Artificial Life*, Varela, J.F. and Bourgine, P. (eds), Cambridge, MA: MIT Press: xi–xviii.

Butler, J. (2006) *Gender Trouble: Feminism and the Subversion of Identity*, New York: Routledge.

Clark, A. (2003) *Natural-born Cyborgs. Minds, Technologies, and the Future of Human Intelligence*, Oxford: Oxford University Press.

Collins, R. and Skover, D. (1992) 'Paratexts', (44) *Stanford Law Review* 3: 509–52.

Davidson, D. (2001) 'Actions, Reasons, and Causes', in idem, *Essays on Actions and Events*, New York: Clarendon Press: 3–20.

Dreyfus, H. L. (1992) *What Computers Still Can't Do: A Critique of Artificial Reason*, Cambridge, MA, London: MIT Press.

Dworkin, R. (1991) *Law's Empire*, Glasgow: Fontana.

Ellul, J. (1967) *The Technological Society*, New York: Vintage Books.

Gallie, W. B. (1956) 'Essentially Contested Concepts', (56) *Proceedings of the Aristotelian Society*: 167–98.

Glenn, H. P. (2004) (2nd ed.) *Legal Traditions of the World*, Oxford: Oxford University Press.

Haraway, D. J. (1991) *Simians, Cyborgs and Women: The Reinvention of Nature*, New York: Routledge.

Hayles, N. K. (1999) *How we Became Posthuman. Virtual Bodies in Cybernetics, Literature, and Informatics*, Chicago: University of Chicago Press.

Heidegger, M. (1977) *The Question Concerning Technology, and Other Essays*, New York: Garland.

Hickman, L. A. (1990) *John Dewey's Pragmatic Technology Indiana Series in the Philosophy of Technology*, Bloomington: Indiana University Press.

Hildebrandt, M. (2008a) 'A Vision of Ambient Law', in *Regulating Technologies*, R. Brownsword and K. Yeung (eds) Oxford: Hart.

—— (2008b) 'Legal and Technological Normativity: More (and Less) than Twin Sisters', (12) *Techné: Journal of the Society for Philosophy and Technology* 3: 169–83.

Hildebrandt, M. and Koops, B.J. (2010) 'The Challenges of Ambient Law and Legal Protection in the Profiling Era', (73) *The Modern Law Review* 3: 428–60.

Horn, P. (2001) 'Autonomic Computing: IBM's Perspective on the State of Information Technology', available at www.research.ibm.com/autonomic/manifesto/autonomic_computing.pdf.

Hutton, P. H., Gutman, H., Martin, L. H. and Foucault, M. (1988) *Technologies of the Self: A Seminar with Michel Foucault*, Amherst: University of Massachusetts Press.

Ihde, D. (1990) *Technology and the Lifeworld: From Garden to Earth*, Bloomington: Indiana University Press.

Ihde, D. (1993) *Philosophy of Technology: An Introduction Paragon Issues in Philosophy.* 1st ed., New York: Paragon House.

ISTAG (2001) *Scenarios for Ambient Intelligence in 2010*, Information Society Technology Advisory Group, available at www.cordis.lu/ist/istag-reports.htm.

ITU (2005) *The Internet of Things*, Geneva: International Telecommunications Union (ITU).

Kranzberg, M. (1986) 'Technology and history: "Kranzberg's Laws"', (27) *Technology and Culture*: 544–60.

Kurzweil, R. (2005) *The Singularity is Near: When Humans Transcend Biology*, New York: Viking.

Latour, B. (1993) *We have Never Been Modern*, Cambridge, MA: Harvard University Press.

Lessig, L. (2006) *Code Version 2.0,* New York: Basic Books.

Morse, S. J. (2008) 'Determinism and the Death of Folk Psychology: Two Challenges to Responsibility from Neuroscience', (9) *Minnesota Journal of Law Science & Technology* 1: 1–35.

Rip, A. (2003) 'Constructing Expertise: In a Third Wave of Science Studies?' (33) *Social Studies of Science* 3: 419–34.

Rip, A., Misa, T.J. and Schot, J. (1995) *Managing Technology in Society: The Approach of Constructive Technology Assessment*, London: Pinter.

Solum, L. B. (1992) 'Legal Personhood for Artificial Intelligences', (70) *North Carolina Law Review* 2: 1231–87.

Verbeek, P.-P. (2005) *What Things do. Philosophical Reflections on Technology, Agency and Design,* Pennsylvania: Pennsylvania State University Press.

—— (2006) 'Materializing Morality. Design Ethics and Technological Mediation', (31) *Science Technology & Human Values* 3: 361–80.

Vismann, C. and Winthrop-Young, G. (2008) *Files: Law and Media Technology*, Stanford: Stanford University Press.

Weiser, M. (1991) 'The computer for the 2st century', (265) *Scientific American* 3: 94–104.

Chapter 1

Smart? Amsterdam urinals and autonomic computing

Don Ihde

When I first learned that the famous fly images in Amsterdam Airport urinals were examples of social-technological "nudging," I was delighted. The example occurs in the book, *Nudge,* by the behavioral economist, Richard Thaler and law professor, Cass Sunstein (now Obama's chief of the Office of Information and Regulatory Affairs) (New York Times 2009).[1] That was because I frequently had earlier used the term "nudging" to indicate a non-deterministic mode of shifting directions or trajectories in human-technology relations. The other frequently used notion I had advised concerning the role of philosophers of technology in relation to social-technological process was that of sitting philosophers in "R & D—research and development" positions in order to have a "place" for "nudging" at design and development stages, rather than after technologies are already developed and "in place" (Ihde 2002: 103–12). Thus, hypothetically, I should be happy to deal with the theme of "autonomic computing," a focal theme for this volume.

"Autonomic Computing," however, becomes highly problematic in precisely this context *because autonomic computing simply does not* (yet. . .and may never?) *exist!* At best, it is a dreamed of Cervantes windmill, which can be tilted against if one is critical, or bypassed in favor of a more empirical or concrete approach. The implied comparison: Amsterdam urinals as a social-technological nudge example, and autonomic computing as a current technofantasy simply are not commensurate. By posing my introduction in this way, I clearly now must explain the problem.

Nudging and the Amsterdam urinals

I begin with the actually existent social-technological example: Amsterdam Airport urinals. These urinals have a fly-image etched onto the porcelain of the urinal, just a bit above the drain outlet. The claim is that once so installed, the spillage on the floor of the men's toilets was reduced *by 80 per cent*! It is thus claimed that this engineered design manages to "attract people's attention and alter their behavior and alter their behavior in a positive way . . . Men evidently like to aim at targets" (New York Times 2009). Now, assuming the claims related to cleaning are correct, and taking note that the Amsterdam Airport is the hub of a very high number of users, and it is also a place where men of many cultures are in transit, then this is a highly successful social-technology. And although on the surface this may appear to be a lo-tech and low-cost technology, its "profiling" and behavior-modifying strategy parallels much of what is desired with hi-tech social-IT strategies especially with respect to marketing.

A second, deeper glance however, shows that the situation is somewhat more complex. Although a fly-image etched simply onto the porcelain remains a simple design, it actually fits into a complex and far-reaching *system* of waste technologies. For instance, far more important even than the clean-up in labor saved – and especially for a country such as the Netherlands where water tables are of essential importance – is the amount of water used to keep the system working. Low-level flushes, coupled to automatic sensors, begin to make the urinal much more hi-tech than might first appear. And these design effects are more "autonomic" than the aim-at-target phenomenon since the user does not directly relate to the flush amount or sensor. Nor is the urinal example as complex as a bi-gender problem, which we New Yorkers are familiar with. In recent years, in theatres, sports complexes and other mass public sites, the phenomenon of women invading men's toilets has produced an amount of amusing commentary equal to that surrounding the literature about the Amsterdam Airport urinals. The lines outside women's toilets are usually much, much longer than those for men and thus in the rush between acts, "invasions" into the men's toilets are known to happen. Today, of course, with the building of the new baseball stadiums, designers have become aware of the need of greater concern, not only for gender ratios, but for time-at-station, and thus the new facilities have proportionately many more female toilet stations than older facilities. (The unisex solutions which exist in many European contexts, would be unlikely to work in the US for obvious cultural reasons.) In short, the Amsterdam Airport urinals are but one instance of a modern sanitation and waste social-technological problem. And this problem encompasses not only the technical and engineering dimensions, but the cultural, social, gender dimensions as well. Even deeper, behind this complex of problems lies the need for critically informed engineering-design education sensitive to the cultural-social as well as the technical demands for design process. And, there remains the issue I shall call "ontological" as well: what shapes the human-social and even "existential" relations to technologies? Now, while I shall return to the issues raised by social-technological "nudging," since it contains provocative ideas about choices and human-technology interactions, I now turn to the problems facing *autonomic computing*.

Autonomic computing

The idea behind autonomic computing is for a totally "autonomous" network system described by Matt Villano (2009) as,

> "a network administrator's dream, a system that identifies, isolates and repairs glitches all by itself. As thresholds are reached, servers reallocate resources

automatically. When a virus or hacker intrudes, the network responds without involving humans at all, eliminating threats and learning from the incidents so they don't occur again. As a result, the technology reduces overall IT cost of ownership by as much as 50 percent."

Here, then, is the dream of a totally autonomous technological system – but more lies behind this dream since such a system is called "autonomic." Turning now to a more serious and technical source, Manish Parashar and Salim Hariri call for a new paradigm to guide the development of autonomic computing,

> "The increasing scale complexity, heterogeneity and dynamism of networks, system and applications have made our computational and information infrastructure brittle, unmanageable and insecure. This has necessitated the investigation of an *alternate paradigm* for system and application design, which is based on strategies *used by biological systems to deal with similar challenges*-a vision that has been referred to as *autonomic computing*."
>
> (Manish and Hariri 2005: 447; emphasis mine)

But which biological system? The answer is the "autonomic nervous system." "The human nervous system is, to the best of our knowledge, the most sophisticated example of autonomic behavior existing in the human body. It is the body's master controller that monitors changes inside and outside the body, integrates sensory inputs, and effects appropriate response. . .the nervous system is able to constantly regulate and maintain homeostasis" (Parashar and Hariri 2005: 248). So, once again in the recent history of computational technologies, the dream of what I shall call *animal autonomy* returns. Before looking at its predecessor, however, I need to expand upon what such a new nervous system paradigm is thought to entail for the designers of its IT analog. I shall here skip to the characteristics deemed desirable for this self-correcting, autonomous system as described by Parashar and Hariri (2005: 255) (in abbreviated form):

> **Self-awareness:** An autonomic application/system "knows itself" and is aware of its state and its behaviors.
> **Self-configuring:** . . .able to configure and reconfigure itself under varying and unpredictable conditions.
> **Self-optimizing:** . . .able to detect suboptimal behaviors and optimize itself to improve its execution.
> **Self-healing:** . . .able to detect and recover from potential problems and. . .function smoothly.
> **Self-protecting:** . . .capable of detecting and protecting its resources from both internal and external attack. . .maintaining. . .security and integrity.
> **Context aware:** An autonomic application/system should be aware of its execution environment and be able to react to changes in the environment.

Open: . . .must function in a heterogeneous world. . .across multiple hardware and software architectures. . .

Anticipatory: . . .able to anticipate to the extend possible, its needs and behaviors and those of its context, and be able to manage itself proactively.

As anyone can see, *this is a very tall order*! But as I read this list and description of the new biological paradigm I also recognized that much of it recalled a similar set of hopes and claims made in a preceding similar utopic battle some decades ago: I refer to the mid-twentieth-century contestations surrounding the then idealized notions for *artificial intelligence* in which the philosophical critic, Hubert Dreyfus, played a central role. This previous contestation produced a "library" of books and articles and spanned argument over several decades. The entire history is too complex to trace here in detail. But I shall look at two of its dimensions which so closely parallel the current state of autonomic computing phenomena. The one dimension may be characterized as the role of *techno-hype* which accompanies and has accompanied in modernity, the designer-corporate claims usually originating from the supporting engineering-corporate developers and then amplified by enthusiastic media-hype in the press and now on the internet. Here the issue is one for a *critical hermeneutics* which must address the mythologization/demythologization issues which surround such technological developments. The second dimension is substantive and relates to the appropriateness of the models and paradigms which are used to frame the development of hoped for technologies. In the earlier artificial intelligence debates Dreyfus highlighted both these dimensions.

I begin with techno-hype: Dreyfus characterized the early period of artificial intelligence (AI) as the days of heady predictions (1957–67) such as those claimed by Herbert Simon. Simon predicted in 1957,

> That within ten years a digital computer will be the world's chess champion [I take preliminary note here of the much later—1957/1997 or forty years later—Big Blue vs. Kasperov match. I will return to this event since I contend it has been terribly misinterpreted and misunderstood!]. . . .within ten years a digital computer will discover and prove an important new mathematical theorem. . .[and] within ten years most theories in psychology will take the form of computer programs, or of qualitative statements about the characteristics of computer programs.
>
> (Dreyfus 1993: 82)

To Simon's predictions were others which included claims that AI processes would produce translation programs, produce a general problem solver, be able to write news stories and the like, all predicated on a belief that AI would exceed human intelligence. Much of this research was then being funded by RAND, MIT, and similar early computational centers. Ironically, by the end of the prediction-decade none of the claims were being fulfilled and it was RAND, worried that the predictions were not coming through, which commissioned Dreyfus to examine the situation which, he and his brother, Stuart, began with the first result, Dreyfus's famous, "Alchemy and Artificial Intelligence" (1967). What I should like to draw attention to here is that these hyper-claims belong to a now very long *tradition* of modernist hopes and projections stemming from initial and often utopian projections concerning new technologies. Moreover, this *myth-making*, I contend, is part of the deep culture of modernist technoscience overall. Its roots go back to the earliest glimmers of modernity, for example, in the Roger Bacon fantasies about underwater boats, flying machines, machines of war (1270) which later were visualized in Leonardo's technical drawings of diving suits, flying machines and human or horse-powered tanks (1450s). Leaping to the twentieth century, recall similar dreams for the elimination of insect born diseases with DDT, almost infinitely cheap electric power with nuclear energy, and the solutions to human hunger with the green revolution. All these overly utopian extrapolations over-looked side effects (insect resistance through mutational change), complications (need for multiple redundancy safety and waste storage systems), and unforeseen hidden costs (transfers of agricultural technologies calling for fertilizer-pesticide-modern technology equipment too expensive for developing countries).

Accompanying, but not fore-fronted here, were also the parallel *dystopian* predictions and mythologies which, in contrarian fashion, project disasters and the destruction of traditional cultural values. These are to my mind, equally exaggerated and absurd. Rather, what I am pointing to is the *prediction pattern* which accompanies each new social-technological development. And what I am advocating is a more "empirical" and historical *critical* look at developing technologies. Such a perspective would reveal that even underway in development, technologies are constantly changing and undergoing modification, often such that when one finally reaches some level of stabilization the technology will rarely look like what the original design or "intent" claimed. Bruno Latour's descriptive history of the Diesel engine is a good example of this phenomenon: although framed by Latour's complex theory concerning the constructions of facts and things, the process by which Diesel's engine becomes a "diesel engine" is suggestive of late modern technology development.

Rudolph Diesel both experimented – one of his predecessor engines using steam exploded and nearly killed him – and theorized. He was enamored of the then current darling of science theory, Carnot's thermodynamic theory. Seeking a more efficient thermodynamic engine, Diesel had ". . .an idea of a perfect engine working according to Carnot's thermodynamic principles. . .an engine where ignition could occur without an increase in temperature, a paradox that Diesel solved by inventing new ways of injecting and burning fuel. . . .we have a book he published and a patent he took out. [patent issued in 1887]" (Latour 1987: 105). Thus, paralleling our autonomic computing story, we have an ideal plan. But, as Latour in his inimical style points out, the process of *materializing* an engine turns out to be more complex and difficult than outlined in the ideal description. Entering into agreements with MAN, Krupp et al., a bevy of engineers gathered now into a

workshop, began to try to produce a diesel engine. "The question of fuel com-
bustion soon turned out to be more problematic, since air and fuel have to be mixed
in a fraction of a second. A solution entailing compressed air injection was found,
but this required huge pumps and new cylinders for the air; the engine became
large and expensive. . ." (Latour 1987: 105). In short, even at this early develop-
mental stage Diesel had already. . . "drifted way from the original patent and from
the principles presented in his book" (Latour 1987: 105).

In any case, ten years later, in 1897, an actual engine had been developed, hope-
fully to be replicated and run by those who buy the license rights to make engines
– but then it turned out the engines were not unproblematic. . . "the engine kept
faltering, stalling, breaking apart. . . .One after another, the licensees returned the
prototypes to Diesel and asked for their money back. . .Diesel went bankrupt and
had a nervous breakdown[1899]" (Latour 1987: 106). Despite these setbacks,
engineers kept tinkering and redesigning, producing engines which could run all
day, but which still needed to be overhauled at night, until 1908 when Diesel's
original patent expired. Then, once in the ". . .public domain, MAN is able to offer
a diesel engine for sale after yet more tinkering, which can be bought as an unproblematic, albeit new, item of equipment. . .[this now is twenty plus years after the
original patent]" (Latour 1987: 106). Although this is a highly foreshortened
history, it is quite typical of technological development as discerned by historians
of technology. The end result turns out to be very different from the initial conception and the way from the initial plan to, in this case a successful, result is
difficult, filled with twists and turns and often the end result does not look anything
like its conceptual beginning. And, even here, I am disregarding the even vaster
history of technologies which fail or do not end up successfully at all.[2] My point
is that a historical and empirically critical analysis must contain skepticism about
techno-hype which remains embedded in technofantasy. Nor is it accidental that
today such hype reflexively points back to the supporters of these vastly expensive
projects. Big Blue was an IBM project and IBM is again behind the autonomic
computing project, with Cisco Systems in a formal partnership to develop an end-
to-end autonomic system, announced in 2004 (Villano 2009).

Turning briefly back to the substantive goals of autonomic computing, however,
also reveals much about some lessons learned from the earlier AI controversy as
well as to lessons *not* learned. For example, three characteristics listed as necessary
for autonomic computing can be seen as directly stemming from the Dreyfus
critique of the formalized, atomistic and essentially closed paradigm belonging to
AI models in mid-century last: Parashar and Hariri argue that an autonomic system
must be *context aware*, and *open*, if it is to be able to perform in parallel with the
autonomic nervous system. These are precisely characteristics which Dreyfus had
claimed the rationalistic AI model of intelligence did not, and could not have –
lessons learned.

Contrarily, Parashar and Hariri clearly have not learned that such systems are
not, in the human sense, *conscious*! So, when they also argue that autonomic computing must be *self-aware, context aware,* and *anticipatory*, they again import into

this paradigm elements of *consciousness*. Yet, if I understand anything about the autonomic nervous system, it is precisely that *it is not conscious*! Homeostasis regarding heartbeat, normal breathing, regulating glucose levels, and all the other "unconscious" adjustments of the autonomic nervous system precisely relieve us of having to consciously adjust our internal stability. Moreover, the kind of consciousness implied in the above computer characteristics are precisely those of an *executive* consciousness which is the kind of consciousness belonging to the long early modern "Cartesian" tradition of mind. Here are lessons *unlearned* and I would argue that these are features which defeat the very notion of autonomous implied in the paradigm. (I would not deny that forms of biofeedback are possible – one can learn to change heartbeat rates and partially modify other normally autonomous processes, but this is precisely an "intrusion" or interruptive modification of those processes which are normally not so modified.)

Thus, both with a demythologized perspective upon the layers of techno-hype, and a brief substantive critique of the new autonomic paradigm, it seems to me unlikely that autonomic computing will turn out to be much like the current projections made for it. Where, then, does this leave us? It is time for a philosophical interlude concerning human-technology inter-relations.

Human-technology inter-relations

Technologies do not invent themselves. . .but, then, neither do humans invent themselves. Implicitly, this way of putting things points to, I argue, an inter-relational ontology. Inverting positively the above observation, if humans "invent" technologies, then in interaction with those newly invented technologies, humans invent (or, re-invent) themselves via the technologies they employ. Ultimately, an inter-relational ontology is a critique of early modern notions of the self, subject/object, and inner/outer distinctions which led to the very notion of an "autonomous" self. Rather, drawing from traditions such as pragmatism and phenomenology, and more recently actor-network-theory, inter-relationality is the recognition that humans already find themselves in a World and only inter-relationally between the World and us, do we discover both ourselves and our worlds. While this inter-relationality is broader than human-technology relations, it can include and focus upon human-technology relations. This is a mutual co-constitutional process. (Although asserted here in briefest from, it is this style of ontology which I have adapted and developed over some forty years of doing philosophy of technology. I shall not here either trace that history, nor re-argue it extensively, rather I shall more immediately return to examples for which inter-relationality can serve analytically.)

Back to Schipol urinals: one could say, urination is a "universal" human practice but in the case of the fly-imaged urinals the practice takes a certain social-technological shape. Human males from a very diverse set of cultural backgrounds apparently – as *Nudge* points out – *like to aim.* Nor are human males unique in this practice which could be characterized as cultural as much as biological, or as Donna Haraway puts it, "naturecultured."[3] Walking my dog, Samurai the Sheba-Inu, he sniffs along the pathway detecting his predecessor dog's urine marks, stops and overlays this mark with his own. No flies here, but much more is involved than a mere relief of a full bladder. We humans, much less olfactorily acute, seem to settle for a visual "target." And not only is this practice multidimensioned, biological-cultural, but it is thus also inter-relational. The "fly" is in-the-world on the urinal and the human male inter-relating, "takes aim." Nor should one forget the "social" in either the dog or human example: those of us who were Explorer Scouts are all familiar with the way in which the group "puts out" the remaining embers of the campfire!

In the incommensurate case of autonomic computing, at first it might seem that since the goal is to have computation be totally autonomous, or totally without human intervention, that inter-relationality might be thought to disappear. That, to me, is yet another index of the probable impossibility of a purely autonomic system. There are *semi*-autonomous or automatic systems, and these work dependably within certain parameters. I shall use the heating system in my vacation house in Vermont as a simple example: during the winter, visits to the house may be less frequent than a month, so to keep the pipes from freezing we leave the oil fired heating system on at a very low temperature and since 1995 when we first lived in this house, we have had no failure in the system *so long as the normal external parameters remained in place.* Even when the electricity goes off (as it frequently does in this remote area), the furnace re-starts itself on the return of power. However, a few years after moving in, a fierce ice storm came bowling down from Canada, snapping off trees, power lines and thus closed off power for some five days. With temperatures hovering around the freezing point, after some days, I headed north. On arrival I found my trusty caretaker had gotten the wood stoves going, the house was well above freezing and not long after, electricity was restored. My point is that here the external parameter threat exceeded the "autonomic" possibility of the system. Similarly, had fuel run out or other support systems failed, the same short-circuit of autonomy would have been the case. *But the same exceeding of parameters applies to the paradigm source: the autonomic nervous system!* Were there to be some serious external attack – an injury to the heart – or even an internal attack – a heart attack, the autonomic nervous system could be overwhelmed and defeated. Similarly, in today's gearing up for cyber-warfare, microwave attacks, stronger and stronger viruses and 'bots would likely overwhelm any dreamed of autonomy from the as yet unrealized paradigm.

Of even greater analytic import here is my deliberate lack of attention to the inter-relational role of humans in relation to the systems, semi-automatic or "fully" automatic. So allow a return to the examples to now take note of the human-technology inter-relations: in the case of my semi-automatic heating system there are a whole series of human-system events needed to keep it going. The fuel tank had to be regularly filled and, unmentioned, the furnace system requires annual maintenance to clean, tune and replace worn parts. Once the power was out, the caretaker intervenes with an alternate system and I intervene to make sure all is going well and restored. And if one goes much farther back, one would have to include the inventors and fabricators of the heating system in the narrative as well, although they now lay farther back and farther away from the events which overcome the autonomy of the heating system. In my older phenomenology of human-technology relations, automatic or semi-automatic systems are in the *background*. Ordinarily they are part of the furniture of a technologically textured world. They do not demand other than occasional attention and thus the human-technology interactions described are somewhat like a "deistic" relationship – a human acts like the deistic god, tinkering, repairing when necessary, but otherwise leaving the system alone. The system, however, is never perpetually on its own.

Ironically, this situation is not too far from the actual and critical – compared to the mythologized – history of Big Blue versus Kasperov. I have described this history much more completely in *Ironic Technics*, but here I only point out that far from Kasperov losing to an autonomous computer with a chess program – Big Blue – he actually lost to a *collection* of human-plus-computer players in that Big Blue was tweaked between matches by its "deistic" keepers who remained out of sight during and after tweaking.[4] I am willing to wager that any autonomic computing will function in the same human-plus computer fashion!

We now have two partially worked out examples: Schipol urinals and (projected) autonomic computing. Even in the incommensurability of a working system and a hypothesized system, there are some symmetries. The Schipol urinals do not work perfectly – if the spillage was cut by 80 per cent, it was not cut by 100 per cent. Perhaps school boys and very old men account for the remainder? But the nudge works in the right direction, saving both labor time and lowering the use of toxic cleaning chemicals in the process. And the nudge suggests yet another contemporary form, now of computer nudging: *profiling.*

Degrees of profiling

Profiling is also planned nudging. I begin with an example which will be familiar to academics: profiling from book readership and sales. Amazom.com has long established its styles of advertising and of using information coming in from sales to profile its products. For example, in 2005 Amazon was awarded a patent for a system which profiles gift recipients which, in turn, raised alarms with consumer advocates who have long been wary of such customer-profiling practices and this one drew particular attention because it was primarily targeted at children. As one critic put it, "Amazon has continued to set the low bar for privacy on the Internet. It's almost no longer a surprise when the company announces some new way to profile people" (Gilbert 2005). Here, however, I shall look at a profiling prac-tice which relates to book sales from an author's perspective. With my most recent book, *Postphenomenology and Technoscience* (2009) Amazon first lists, "Frequently Bought Together. . ." which turns out to be a list of several other of my books, then under, "Customers who bought this item also bought. . .," and again half a dozen of my previous books are listed. However, when pulling up *Listening and Voice: Phenomenologies of Sound* (2007 Second Edition), in the first category is Nancy's *Listening,* and in the second some eight other books on sound, music and noise with only some authors previously known to me, and from a variety of disciplines. My third instance, *Ironic Technics* (2008) again followed the pattern of listing other books of mine in the first category, but in the second it included *We Have Never been Modern,* by Bruno Latour, *Natural Born Cyborgs,* by Andy Clark, and a surprise, *My Stroke of Insight,* by Jill Taylor. Although authors attending to technologies (such as Latour, in part Andy Clark, to which can be added Andrew Feenberg, Donna Haraway, Kathrine Hayles, and many others concerned with technoscience) appear scattered through the listings of many of my Amazon advertised books, the surprise, Jill Taylor, a neurologist who had a stroke and wrote about it, while known to me, appeared as a new and unexpected entry. Although I shall not cite the examples from my other listings, what first stands out is that there are few surprises, and furthermore, most of the listings are "in the field" in some broad way. Of course, as an author, I should be expected to know these other listings – but for a new buyer or shopper such listings might be mildly informative. I would contend at this point that this low-level profiling is mostly a kind of "insider" profiling and for neither the author nor the shopper would it likely lead to innovative new directions precisely because it is, as insider, relying upon already set interests and patterns. This profiling is "conservative" in that it presupposes set interests and habits of buying.

A different direction can be illustrated with the book, *Nudge* (2008) by Thaler and Sunstein cited above. Although, since this book is written about technological design, its disciplinary origins are behavioral economics and law and a glace at their index shows clearly that these authors *do not cite philosophers, or philosophers of technology.* It is "outside the usual technoscience box." It was Evan Selinger who called my attention to *Nudge* since he had run across and reviewed that book – for me, he was calling to mind his own chapter in *Postphenomenology: A Critical Companion to Ihde* (which he also edited) and his chapter, "Normative Phenomenology: Reflections on Ihde's Significant Nudging" (Selinger 2006). (I had used this concept in *AI and Society* as early as 1999, then reprinted in *Bodies in Technology* 2002.) While, in spite of the somewhat cozy form of this example, Amazon's profiling could have, but did not pick up this connection. One more example from a non-profiling source: as a frequent reviewer for manuscripts for various presses, reviewers are often allowed to choose books from the press's listings either in lieu of an honorarium, or in addition to one. Here the reviewer pulls up the listings and makes choices – in my case I have found this an excellent opportunity to get books of interest "outside the profiling box" and here I may indulge interests well outside the range of past published work which sets the parameter for distributors' profiles. All of these limited examples point to a certain malleability within nudge-type profiling. It remains interactive, of course, because the profiles suggest rather than either demand or restrict. This is a soft sell, not unlike the soft role of the fly-image.

A much more extreme profiling inhabits a *restrictive* profiling, two examples of which are dramatically contemporary. Both examples involve governmental attempts to restrict, censor, eliminate various streams of information for users of the internet. The first example is the current attempt of the Chinese government to restrict *incoming* information on the internet. The project called Green Dam-Youth Escort was a ruling that all computers sold from July 1 must have software which blocks pornography sites (and other objectionable sites, presumably regarding Tibet, Falun Gong and other topics anathema to the Government). As of this writing, a delay has been given in response to protests by manufacturers concerning free trade laws and the international civil and free speech rights communities protesting censorship. This called for blockage, profiling disapproved topics, aims at incoming information.

The opposite blockage has been attempted by the Iranian theocratic government. During the recent 2009 election, highly contested within Iran, the government attempted to close down *outgoing* information. With primarily "centralized" media such as news services, newspapers and even television, the blockage was highly successful. However, "decentralized" and new media, including cell phone photography, distributed internet sources, twitter, blogs, face-book outlets could not be damped down. In the end, a resort to massive police, militia and other classical repressive power tactics have for the moment prevailed. (It is interesting to note that during the Iranian Revolution of 1979, the same success of decentralized, then new media also were used to affect the outcome, i.e., portable cassette tape recorders were the technologies of choice to disseminate information from the anti-Shah opponents!) Both these examples are of *hard* profiling.

Note that in both cases, China and Iran, the new and electronic media not only are hard to suppress, but defeasibility strategies utilized by smart technically informed users easily go around the firewalls, blocking devices and other strategies used by the ruling powers. Thus even with severe negative "profiling" there is leakage, and this leakage follows the pattern of human-technology interaction. Contrarily, nothing "automatic" works or remains impermeable. At the current stage, leakage has become easier and more pervasive precisely because of the shift to "wireless" electronic technologies. The nineteenth-century wired communication devices – telegraph, telephone, cable, etc. – could be severed. To defeat today's wireless communications one would have to interrupt whole systems of transmission, including satellites, microwave towers, disks and other broadcast and receiving technologies, the recognition of which has led to predictable escalation now being called "cyber-wars." Lasers, microwave attack devices, and a plethora of mostly not-yet developed cyber-attack technologies are planned for defeating the various forms of leakage or defeasibility technologies. There is also a detectable pattern to the new contestations. In many instances miniaturized, decentralized and often "cheaper" technologies may be used to defeat much larger, complex and centralized technologies. For example, laser pointers can disable surveillance cameras, and in the long-lasting wars between traffic control police using radar, laser measurement devices or other speed registering instrumentation to detect speeders are countered by miniaturized jammers, detectors and warning devices employed by drivers to defeat the detectors.

Does this same pattern now get reflected in twenty-first century "wars"? I suggest that the answer is affirmative in that many forms of terrorism display precisely the David/Goliath technology pattern here noted. Small hand-held anti-aircraft missile launchers can and have shot down large and expensive helicopters. Improvised roadside bombs defeat armored personnel carriers. And although not miniaturized, mid-size airliners destroy hi-rise skyscrapers. Now while the small, decentralized, and cheaper David-technologies have become weapons in terrorist wars, this is not a matter of more primitive or lo-tech devices employed against more contemporary hi-tech devices. Both the David and the Goliath technologies tend to be hi-tech. Multipurpose cell phones (communication, images, texts, and even detonator capacities), heat sensing missiles both small and large, and robotics all enter the contemporary contestation. Now, including IT technologies, the single or small number of hackers can also invade and sometimes damage vast network systems such as military, financial, or other thought-to-be secure complexes along with identity theft sometimes on a massive scale. I would contend that this pattern now obtains across a very large spectrum of human-technological-social actions and reflects a certain *style* of human-technology inter-relationality appropriate to its contemporary gestalt.

If I am correct about this pattern, then it looks very different from the early twentieth century worries of the first generation of philosophers of technology. These early worries were directed at gigantism, at a tendency perceived as destructive of high culture and the enhancement of mass man or the coming of a look-alike common man, and worries about centralized and authoritarian regimes. Echoes of these worries may be found in Jacques Ellul, Martin Heidegger, Lewis Mumford, and the first-generation Frankfurt School thinkers. If anything, the David/Goliath pattern *inverts* the earlier and largely industrial paradigm.

Multitasking and multistability

The narrative above implicitly follows what has been called the "empirical turn" in the philosophy of technology. By looking at examples, demythologizing from technofantasies, and yet still trying to discern patterns, what emerges looks on the surface like a more empirical, but also "messy" display. The mess, however, is not so much mess as a display of a deep possibility-pattern of human-technology interactions which I have termed, *multistability*. Multistability is the variational range which appears – and I would say always appears – in the uses or practices which humans using technologies may invent and learn. Let us begin with tool-technologies: a tool, unused, might be simply an "object" lying around or neatly placed in a tool set and, contrarily, a human without his/her tool finds himself/herself only able to perform tasks limited to bodily capacities. But, with a hammer – here I deliberately follow the long tradition begun by Heidegger – human-tool actions multiply.

Heidegger did not specify, in his famous example, what kind of hammer he was referring to (hammers remain undifferentiated just as Dasein is ungendered, in typical Heidegger "metaphysical" style). So, let us take a claw hammer, designed for carpenter use. Its head is flat for driving nails, its claw curved for pulling them out. Yet, I would contend, *no technology ever remains limited to "designer intent."* And while Heidegger's analysis recognizes that a tool is what it is in relation to its context of uses, he was less sensitive to the multiplicity of uses which could (and does) arise from hammer practices.[5] Beginning with a trajectory from designer intent, the head or pounding end may be used in any number of ways thus making, in use, the hammer into a "different" technology, or better, revealing a different human-technology action/interaction. Some years ago a disturbed ex-graduate math student used a hammer to murder one of his ex-professors in Berkeley, California. Hammer becomes weapon. Or, our own Stony Brook philosophy grad students imaginatively began to decorate their doors as "art works." Each previously brown, wood door was decorated with different displays, sometimes painted or covered (one was covered with false fur), sometimes containing *objets d'art.* One of these had half a dozen hammers screwed onto it: claw hammer, ball pein hammer, tack hammer, etc. Here the hammer becomes conceptual art, to which one may add any number of other imagined or actual uses such as hammer as projectile (thrown), window opening device (when propped into a window), finger or toe cracker (misaimed or dropped) *ad infinitum.* The hammer, I have used because in its con-servative designer intent, is a very simple tool-technology. Contrarily, multi-uses may also be designed into tool-technologies, for example, the Swiss army knife.

The Swiss army knife is designedly a *multitasking* tool with several blades, a bottle and can opener, corkscrew, to which are sometimes added small scissors, screwdriver blade, picks and the like. I always used to carry one with me on travels, but since 9/11 these are banned. Its electronic offspring is the cell (or mobile) phone, also designedly a multitasking technology. Communication device, both voice and text, camera, sometimes bar code reader and recorder of purchases, alarm clock, and as above, detonating device (cell phones are still allowed for travelers, post-9/11, but could be used for much more deadly purposes than Swiss army knives!). The implication, however, is that all tool-technologies, even all technolo-gies in human-technology inter-relations are *multistable.* Designed multitasking simply materially recognizes this possibility structure. There are two obvious implications which may be noted: first, the variability which arises, belongs to human-technology interactivity. Left on a shelf, the Swiss army knife or the cell phone "does" nothing. And without the technology, the human is limited to his/her bodily actional capacities, but inter-relationally the multiple trajectories and uses proliferate. Second, this makes technological *predictions* very, very difficult!

Urinals and autonomic computing redux

It is now time to return to the thematic examples which began this narrative. First, the Schipol urinals: a confession, the first time I used the fly-imaged airport urinal, I was amused by the image but did not think of it as a target (although I also did not "spill"). I have previously argued that technologies, including the soft nudge types, are not *deterministic* but do display over a vast number of users and multiple events that there is something like a center-of-gravity or an *inclination* within the trajectory presented. This compliments the notion of a nudge technology. However, it does not determine how the human-technology interaction must take shape. Thus, for either inexperienced or unable users, as suggested above, the unintended (design intent) spillage can still and does occur. Then hypothesize a knowing user, one who discovers he is being nudged and rebels against the nudge – a deliberate spiller reacting in a double hermeneutic action against the intended design use. All of this lies within the maximal range of human-technology interactions and points to the inherent limits to anything like technological control. And when such exceeding of limits occurs, then in situations where authorities want control, much more repressive forces may be used. While the current Iranian authorities could not prevent information leakage, and wherein this leakage prompted a greater international sympathy with the election opponents, the repressive use of police, militia and in current news, the arrest and torture of opponents to gain public confessions can be said (temporarily?) to trump leakage.

Returning to autonomic computing, the development if guided by the present "new paradigm" cited, wants to have it both ways. On the one hand, by becoming purely autonomic or autonomous, the human-technology interaction is wishfully excluded. I have suggested that this is impossible and particularly in situations which exceed the internal limitations of the autonomic system, just as with the autonomic nervous system, the autonomy can be breached perhaps not as easily as the rebellious urinator, but breached nonetheless. On the other hand, the current autonomic technofantasy wants to make it easier for the "controllers" lowering the amount of human intervention or human-technology interactivity needed to run smoothly and cheaply. The latest "Hal fantasy (*2001*)" is revived again,[6] benignly, in Peter Hughes in NASA's Information Systems Division. Hopeful that the IBM end-to-end system can produce computers which fix themselves, he says: "If a failure occurs in space, we want to be able to rest assured that the system can correct itself without intervention of any kind. . .on the business level, you can always throw humans at a problem in a pinch, in space, it's a different situation entirely" (Villano 2009). My suspicion is that just as NASA has had to "throw humans at a problem" from Apollo 13 to the Mars Explorer, this continuation of human-technology interactivity will continue.

Notes

1 The original discussion may be found in Thaler and Sunstein (2008: 3–4).
2 Bruno Latour (1996) traces the planning and ultimate failure of an individualized railway system. Similarly, John Law (2002) traces the even more expensive development and planning of a Cold War fighter bomber, the TSR2, which was cancelled. Detailed histories of failed technologies are rare despite the fact that many projected technologies do fail.
3 This term and its counterpart, culturenatures, are common to Donna Haraway's most recent books, including Haraway (1997).

4   Ihde (2008), see especially chapter four, 'Of which human are we post?' which con-
    tains a detailed description of the Big Blue-Kasperov contest.
5   Heidegger (1962: 69 ff). The hammer example lies within Heidegger's discussion of
    ready-to-hand and presence-at-hand.
6   2001 refers to the Stanley Kubrick's famous '2001: A Space Odyssey' released in 1968.

References

Dreyfus, H. (1967) *Alchemy and Artificial Intelligence*, Santa Monica, CA: The Rand
    Corporation.
—— (1993) *What Computers Still Can't Do*, Cambridge, MA: MIT Press.
Gilbert, A. (2005) 'Privacy advocates frown on Amazon snooping plan,' CNET News,
    available http: http://news.cnet.com/Privacy-advocates-frown-on-Amazon-snooping-
    plan/2100–1038_3–5611663.html?tag=mncol (accessed June 2009).
Haraway, D. (1997) *Modest Wisness*, London: Routledge.
Heidegger, M. (1962) *Being and Time,* New York: Harper and Row.
Ihde, D. (2002) *Bodies in Technology*, Minneapolis, MN: University of Minnesota Press.
—— (2008) *Ironic Technics*, Copenhagen: Automatic Press.
Latour, B. (1987) *Science in Action,* Cambridge, MA: Harvard University Press.
—— (1996) *Aramus, or the Love of Technology*, Cambridge, MA: Harvard University Press.
Law, J. (2002) *Aircraft Stories: Decentering the Object in Technoscience*, Durham, NC:
    Duke University Press.
New York Times (2009) *New York Times*, Business, February 28, 2009.
Parashar, M. & Hariri, S. (2005) 'Autonomic computing: an overview,' Unconventional
    Programming Paradigms: International Workshop UPP 2004, Le Mont Saint Michel,
    France, September 15–17, 2004, Springer: 247–59.
Selinger, E. (2006) 'Normative phenomenology: reflections on Ihde's significant nudging',
    in: Selinger, E. (ed.) *Postphenomenology: A Critical Companion to Ihde*, Albany, NY:
    SUNY Press: 89–108.
Thaler R. & Sunstein, C. (2008) *Nudge: Improving Decisions about Health, Wealth, and
    Happiness*, New Haven, CT: Yale University Press.
Villano, M. (2009) *Autonomic Computing: Fantasy or Reality?* CRN News, available http:
    www.crn.com/it-channel/18825106 (accessed June 22, 2009).

Chapter 2

Subject to technology

On autonomic computing and human autonomy[1]

Peter-Paul Verbeek

Introduction

Without too much exaggeration it can be claimed that, ever since the Enlightenment, autonomy has been the crown jewel of humanity. It is not without irony, therefore, that precisely the concept of autonomy has come to be used to indicate the activities of technologies that have a rather problematic relation with human autonomy. Autonomic computing, and its manifestations of ambient intelligence and persuasive technologies, constitute a new generation of technologies that interact intelligently and relatively independently with human beings (cf. Verbeek 2009). Often without people explicitly noticing it, such technologies help to shape human actions and experiences, ranging from automatic systems in cars that overrule driver decisions to persuasive mirrors in medical practices that try to persuade people to develop a healthier lifestyle by predicting how they would look in ten years' time (cf. Fogg 2003).

This chapter aims to address the ethics of autonomic computing by analyzing the conflict between these two forms of autonomy. In doing so, I will not only address how autonomic computing affects human autonomy, I will also focus on the question of the value of the concept of autonomy in ethics itself. Even though the figure of the autonomous subject has acquired a central role in contemporary ethical theory, I will elaborate how this figure eclipses the fundamental role of technological mediation in our technological culture. Rather than as autonomous beings that need to be defended against technology, human subjects need to be seen as technologically mediated subjects. I will defend the thesis that the ethics of technology in general, and the ethics of autonomic computing in particular, needs to focus on the responsible shaping of our technologically mediated subjectivity, and on accompanying the technological developments that form the basis of these mediations.

New relationships between human and technology

Autonomic computing involves a radically new relation between human beings and technologies, that requires further conceptualization. In recent decades, the philosophy of technology has shifted its focus from laying bare the detrimental effects of 'autonomous technology' (Ellul 1964) to identifying and analyzing the relationships between humans and technologies. In this transition, the work of the North-American philosopher Don Ihde played a central role. Starting from the phenomenological idea that human existence can be understood only in terms of our relationship to reality, Ihde (1990) has researched into the many ways in which this relationship is actually mediated by technology. People can *embody* technologies, as when wearing a pair of glasses which one does not look *at* but looks *through*. Other technologies we have to *read*, in the way that a thermometer gives information on temperature or an ultrasound machine gives a representation of an unborn child. People can also *interact* with technology, as when operating a DVD player or setting a central heating thermostat. Finally, within the framework sketched by Ihde, technologies can also play a role in the *background* of our experience. The fan noise made by a computer and the illumination provided by room lights are not experienced directly, but form a context within which people experience reality. Ihde's work has comprehensively researched how technology, mediated by the different relationships that people can have with it, plays a role in the establishment of interpretative frameworks, scientific knowledge, and cultural practices.

Ihde's framework has been of considerable value to the contemporary philosophy of technology. Yet, current technological developments – like ambient intelligence and autonomic computing, or the new possibilities opened up by the ongoing convergence of nanotechnology, biotechnology, information technology, and cognitive science – move beyond the boundaries of this framework. The central focus of Ihde's schema is technology which gets *used*: glasses, telescopes, hammers, and hearing aids. However, the newest technologies increasingly organize man-machine relationships that can no longer be characterized as 'use' configurations.

Autonomic computing and Ambient Intelligence, for example, lead to a configuration that one might rather give the name of *immersion*: here, people are immersed in 'smart environments' that react intelligently to their presence and activities. These technologies are beyond what Ihde calls a 'background relation' with people; because they explicitly engage in interaction with them, they are more than just a 'context'.

An entirely opposite route is taken by the 'anthropotechnologies', to use a term coined by Peter Sloterdijk (1999): technologies which redesign human beings at the physical level, like neuro-implants, genetic modifications, and cognition-enhancing drugs. These technologies are not of the exterior, the environment, but of the interior – within the human body. Our relationship with them goes beyond that of *embodiment*; it might rather be said to represent a *merge*, as it becomes difficult to draw a distinction between the human and the technological. When a deaf person is given a degree of hearing capability thanks to a cochlear implant connected directly to their auditory nerve, then this 'hearing' is a joint activity of the human and the technological; it is the configuration as a whole that 'hears', and not a human being whose 'hearing' is restored thanks to technology (cf. Verbeek 2008a).

Autonomy: the limit of humanity?

Both these technological trends – outwards, towards the environment, and inwards, towards the body – are blurring the borderline between humans and technology. They are also making technology increasingly invisible: it does its work without allowing us to adopt an explicit relationship to it. And this is undoubtedly one of the reasons that some people see the current convergence of technological domains as a potential threat. When smart environments start meddling with us of their own accord, secretly persuading us to change our behaviour, tracking and monitoring our actions on the internet, and registering where we find ourselves at what times, it feels as if we are losing our grip on what happens to us. Our boundaries appear to evaporate: externally, in our environments, and internally, within our own bodies, it seems that technologies are running the show. A living room that decides independently how warm it should be, what colour the lighting should be, and whether the phone is allowed to ring is reducing our autonomy considerably; and the same is unquestionably true of brain implants that mitigate the symptoms of Parkinson's disease but which also bring about personality changes.

When the boundary between the human and the technological is blurred, we also appear to have to give up that which makes us most human: our autonomy, the freedom to organize our lives as we see fit. After all, without this autonomy we are but slaves to technology. A world in which people are directed by devices which do their work invisibly, whether in the environment or from within the body, perfectly embodies the *Brave New World* dystopia that is so widely feared.

It is no exaggeration to say that the relationship between technological power and human autonomy has been an obsession of classical philosophy of technology. From Lewis Mumford's *Megamachine* to Charlie Chaplin's *Modern Times*, the core theme has been: how are we to escape from the dominance of technology? How are we to prevent technology from taking power over people and thereby alienating them from themselves and their surroundings? The reality, however, is considerably more complex. In actual fact we have never been autonomous with regard to technology, not even with regard to technologies we simply 'use' and which are *not* concealed in the environment or within our bodies.

One of the most important insights to have emerged from contemporary approaches to the philosophy of technology is the realization that technology plays a fundamental mediating role in human experience and activity. Our personal contacts are mediated by telephones and computers; our opinions and ideas are mediated by newspapers, televisions and computer screens; and our movements are mediated by cars, trains and aeroplanes. Technology has even played a crucial role in the ethical domain, as I have elaborated in recent years. The decision on whether a pregnancy should be terminated if the child has a genetic disorder, for instance, is not an autonomous choice; to an important degree it is prestructured by the way a modern technology such as ultrasound scanning presents the unborn child (Verbeek 2008c). We must give up the idea that we exercise a sovereign authority over technology and that we employ technologies merely as neutral means towards ends that have been autonomously determined. The truth is that we are profoundly technologically mediated beings.

For modern people like ourselves, however, the product of the Enlightenment, this fact is rather hard to swallow. After all, the modern self-image of the autonomous subject, freed by the Enlightenment from dictatorship, ignorance and dependence, has already suffered some serious dents, as Freud's *A General Introduction to Psychoanalysis* showed all too clearly. Copernicus evicted us from the centre of the universe by having the Earth rotate around the sun; then Darwin took away our unique position in Creation by linking humans to other animals through evolution. Finally, Freud took responsibility for dealing the third blow to our modern self-image by showing that the ego, far from being its own master, is itself the product of a complex interaction with the subconscious (Freud 1989).

Today's technological developments continue to unmask the modern autonomous subject, but by other means than has philosophy. Freud's list of unmaskers of the modern subject was composed entirely of thinkers who showed that we should try to *understand* people in a different way. By now, it needs to be expanded to include a series of scientists who have questioned human autonomy in different ways again. And on this list should definitely be the brains behind *Ambient Intelligence* and *Autonomic Computing* (cf. Aarts and Marzano 2003).

The human condition

The fundamental interconnectedness of humans and technology means that 'the human condition' is not a constant factor to which we could ethically appeal. What makes us human, both in the existential sense and in the biological sense, is *historic*. It has become what it is now, and it will continue to develop. This historic, rather than essentialist character of the human condition has profound consequences. It means that none of the central dimensions of our human existence – like our natality, mortality, freedom and intentionality, but also our appearance and gender – will remain the same forever.

Pre-implantation diagnostics, for instance, makes it possible to prevent the development of embryos with certain genetic properties. Quite apart from the ethical question of whether the application of this technology is desirable, it is clear that human natality is changed by the availability of this technology. To bring a child into the world who carries certain hereditary traits suddenly becomes something for which people can take personal responsibility. In fact, in extreme cases people could even be *held responsible* for it, as in the so-called 'wrongful life' lawsuits in which children sue their doctors, or even their parents, for the fact that they were born at all.

The same applies to our mortality. New technological developments in the areas of palliative care, euthanasia, and intensive care mean that mortality today is not what it was for previous generations. The end of our life is no longer something that we simply undergo, but something we have to make choices about. This is independent of any moral judgement about the desirability of technological intervention at the end of life; the simple fact of the availability of these technologies means that we become responsible.

Even human freedom and intentionality – often seen as the crown jewels of humanity – are subject to continuous technological change. And especially these two aspects of the human condition are at stake in autonomic computing. When people adapt their lifestyle because a Persuasive Mirror confronts them with the potential consequences of their behaviour, a decision to change one's life is not a fully autonomous decision. In such cases, human beings have allowed themselves to be educated by technology, and their intentions have become interwoven with those of technology. Such a mirror does not simply condition human behaviour; it rather helps to shape the interpretations on the basis of which human beings make intentional decisions. Intentionality has taken on a hybrid character here: it is partly human and partly non-human. Subjects who act and take decisions in these cases, do not do so as purely 'human' beings, but as human-technology hybrids. Decision-making appears to have become a joint affair, concerning both humans and technology.

Blurring boundaries

What good does it do to equate today's blurring of the border between humans and technology with the unmasking of the 'autonomous subject'? Does this approach leave us no option than to simply accept that we are slaves to technology, free only to display the occasional bout of subversive behaviour? Can we even talk about ethical limits to technology if our minds and bodies are entirely mediated and directed by that technology? Must we simply accept that the border between humans and technology is a fiction, and deliver ourselves to the machines?

No, of course we must not. Precisely that would actually mean the end of humanity. The diagnosis that humankind is controlled by technology, and that no more than token subversive resistance can be offered, fails to appreciate how each is interwoven with the other. And in doing so, it also loses the possibility to take responsibility for the quality of this interweaving. There is an interplay between humans and technologies within which neither technological development nor humans has autonomy. Humankind is a product of technology, just as technology is a product of humankind.

This does not mean that we are the hapless victims of technology, though; but neither does it mean that we should try to escape from its influence. In contrast to such a *dialectic* approach, which sees the relationship between humans and technology in terms of oppression and liberation, we need a *hermeneutic* approach. Within such an approach – hermeneutics is the study of meaning and interpretation – technology forms the tissue of meaning within which our existence takes shape. We are as autonomous with regard to technology as we are with regard to language, oxygen, or gravity. It is absurd to think that we can rid ourselves of this dependency, because we would remove ourselves in the process. Technology is part of the human condition. We must learn to live with it – in every sense of the word. In other words, we must shape our existence *in relation to* technology.

And this is where we encounter a metaphysical issue which forms the crux of the philosophy of technology. At the source of the dialectical approach to the philosophy of technology, and its narrative of oppression versus liberation, lies a very specific metaphysical concept of the relationship between humankind and reality. As the French philosopher Bruno Latour has argued, this concept, which has characterized all of post-Enlightenment modernism, draws a fundamental distinction between 'subjects' and 'objects'. Subjects are active, have intentionality and freedom; objects are lifeless, passive, and at best serve as the projections or instruments of human intentions (Latour 1991). Such a metaphysics makes it impossible to properly discern the interrelatedness and interconnectedness of subject and object – of humankind and technology. The moral significance of technology, the technologically mediated character of human freedom, and all the ways that people express their humanity through relationships with technology – all of this is rendered invisible by a modernistic metaphysics which radically separates subjects and objects and diametrically opposes them.

However, what has hitherto remained absent in a non-modernistic or a modernistic perspective of the type proposed by Latour is an ethics to replace the unilateral rejection that is characteristic of the classic critique of technology. We must not only develop a concept of humankind which goes beyond the 'autonomous subject' that wants to be purged of all outside influence, but we must also develop an ethics that goes beyond safeguarding this purging and which looks further than the risks, the violations of privacy, and the other threats that 'technology' poses to 'humanity'.

Ethics

Toward a non-humanistic ethics

The analysis I have given so far, of an increasingly blurred borderline between humans and technology, might give the impression of being entirely ethically nihilistic. After all, if no real borderline can be drawn between humans and technology, and if we never were as autonomous and authentic as we thought, then what's the use of ethics? If technology mediates our whole existence, from birth to death and everything in between, then why would we trouble to look at technology through the lens of ethics?

If this was your impression, then I am happy to say that I can reassure you. In my view, the analysis I have presented so far, framed as it is by the philosophy of technology and philosophical anthropology, only really comes into its own as a contribution to the ethics of technology. Putting the borderline between people and technology into perspective certainly does not mean that from now on 'anything goes'. On the contrary: it means that the aim of the ethics of technology must be to give shape, in a sound and responsible way, to the relationship between people and technology.

This is going to be no simple matter, however; today's ethics of technology leaves much to be desired. It is dominated by what I have called an 'externalistic' approach towards technology (Verbeek forthcoming 2011). The basic model is that there are two spheres, one of humanity and one of technology, and that it is the task of ethics to ensure that technology does not transgress too far into the human sphere. To stay within the paradigm of the 'limits of humanity', in this model ethics is a border guard whose job it is to prevent an unwanted invasion. However, in the light of the analysis I have presented here of the relationship between humans and technology, this model is inadequate; it draws a distinction between a 'human' domain and a 'technological' domain which is ultimately untenable. And while smart environments, brain implants, and embryo selection have already begun their advance, this ethics is painting itself into a corner by only being willing to consider the question of whether such technological developments are morally acceptable or not.

So in ethics, too, we must cross the boundary between subject and object. We must no longer see ethics as a matter concerning the subject alone, but as a coproduction of subject and object. Over recent years I have elaborated one possible direction for such an 'amodern' ethics by researching into the moral dimensions of technology. The example I just gave, of the moral significance of ultrasound technologies, has formed a guiding example; the ethical decisions surrounding abortion cannot be seen as an autonomous moral human choice, because they are formed, to an important degree, by the way that technologies like ultrasound present the unborn child.

However, as I have explained, the technological developments at the heart of this address have regard to another configuration than that of the *use* of technology, and in so doing depict a new form of the interrelatedness of human subjects and technological objects. In the *blending* and *immersion* configurations, our lives get shaped in radically novel ways. And the ethical questions here are considerably more delicate.

For the configuration of *blending* this was well illustrated by the furore that arose ten years ago after Peter Sloterdijk gave his renowned speech *Rules for the Human Zoo* (Sloterdijk 1999). In this speech, Sloterdijk argued that the latest technologies offer entirely different media by which we could give shape to our humanity, media other than those of the word. While texts had always been used to *tame* people, new technologies were making it possible to *breed* them, and according to Sloterdijk it was high time to start pointing these new possibilities in the right direction. But while philosophers racked their brains about the texts and ideas that formed people, the actual material re-creation of humanity was proceeding apace. In a provocative formulation, Sloterdijk proposed that 'rules for the human zoo' were needed; people did not live merely as conscious minds in a universe of ideas, but also as organic beings in a biotope – a 'zoo' – and it was this organic dimension of our existence which now needed our full attention.

The German academic world was in uproar after this speech. Sloterdijk's plea that rules should be developed for human 'cultivation' was immediately associated with Nazi eugenics programmes. Jürgen Habermas, for instance, who according to reports was active behind the scenes in the attack on Sloterdijk, has since published a book in which he explicitly states that genetic intervention should be allowed only for *therapeutic* purposes: all interventions aimed at human *enhancement*, such as pre-implantation genetic diagnosis and genetic enhancements, are morally unacceptable, because these technologies mean that we take decisions *on behalf of others* about what kind of life is worth living (Habermas 2003). By erasing the difference between the 'grown' and the 'made', these technologies attack the 'autonomous authorship of existence' and the 'moral self-understanding' of the person so 'programmed' (idem, 52). Today's anthropotechnologies treat people not as self-actualizing subjects, but as the 'instruments of our preferences' – and this Habermas finds utterly unacceptable.

Simply posing the question of how best to shape the interrelatedness of humans and technology, then, had turned out to be rather too much of a good thing. But while intellectuals struggled to outdo each others' political correctness and proclamations on the evils of eugenics, the ethical questions stood, and remained unanswered. This was a clear instance of the failure of the modernistic perspective on ethics; while in the real world humans and technologies are becoming ever more intertwined, ethics stands on the sidelines, hawking a division of the estate.

This dispute between Sloterdijk and Habermas can easily be transposed from the realm of biotechnology to autonomic computing and ambient intelligence. As we saw, just like more biologically oriented forms of 'anthropotechnology', advanced forms of information technology have the potential to change what it means to be a human being. And most often this implies that also in the field of autonomic computing designers make decisions on behalf of others about what kind of life is worth living. The persuasive mirror mentioned above embodies implicitly embodies the norm that it is better to live a healthy and moderate life than a more intense life with a higher risk of disease. Ambient intelligence in health care implicitly reconfigures the lives of people in need of care. The introduction of smart walls ad smart beds in geriatric hospitals, for instance – that can detect if people are shouting for help or fall from their bed – will reduce the contact patients have with nurses. Behind our backs, technologies change what it means to be a human being – and we need to face the question of what kind of ethics can be able to guide these developments.

Beyond ethics as boundary guard

To be sure: I share the belief that we should respect the rights of others as far as possible, and that we should treat people as ends in themselves and not as means to an end. Actually, there must be very few people in this society who do *not* share this view. But at the same time, the philosophy of technological mediation shows that it is a fiction to suppose that a society is imaginable in which people can take entirely autonomous decisions on what kind of life is worth living. The remarkable thing about technology is that it contributes *continuously* to the way we answer questions about the good life. Autonomic computing has simply added new components to an existing repertoire.

Yet, this does not mean that we are incapable of dealing responsibly with them. Instead of making ethics a border guard who decides the extent to which technological objects may be allowed to enter the world of human subjects, ethics should be directed towards the *quality of the interaction* between humans and technology. This does not mean that every form of such interaction is desirable, nor that we should simply develop technologies at random. I agree with Habermas that respect for individual freedom and for human dignity must play an important role in this matter. But the distinction between 'therapy' and 'enhancement' fails to provide an appropriate vantage point. We cannot employ the criterion that we must stop at the point where the 'restoration' of an original situation gives way to the creation of a new human being; after all, the 'original situation' does not exist, and we have always used technology to create ourselves anew.

Every technological development puts at stake what it means to be a human being. The anticonceptive pill has dramatically changed our experience of sexuality, and the disconnection it affords between sexuality and possible parenthood has played an important role in the emancipation of women and gay couples. The possibilities offered by anaesthesia and palliative care have played an important role in our current views of what constitutes a dignified life and a dignified death. In its own distinct way, autonomic computing will reshape what it means to be a human being as well, by allowing us to delegate aspects of our freedom, intentionality, and even morality to intelligent technological systems and devices.

The crucial question here is not so much where we have to draw the line – for humans, or for technologies – but how we are best to shape the interrelatedness between humans and technology that has always been a hallmark of the human condition. Now that we have started to trust parts of ourselves to autonomic computing, what could be good ways to do this? We need an ethics that does not stare blindly at the issue of whether a given technology is morally acceptable or not, but which looks at the quality of life as lived with technology. I should like to conclude this chapter with a proposal for just such an ethics.

Governing the mediated subject

The good life

In elaborating a non-modernistic ethics it is useful to connect to an approach that originated in classical antiquity – and which was obviously, and by definition, non-modernistic. At the core of classical ethics was the concept of 'the good life'. Ethics had not so much to do with the question of 'how I should behave', as a moral subject in a world of objects, but with the question of *how to live*. The good life was directed by *aretè* – a term frequently translated as 'virtue', but which is better rendered by the word 'excellence'. Ethics, then, was about mastering the art of living.

A very inspiring reading of this 'ethics of the good life' is offered by Michel Foucault's work on ethics (Foucault 1990, 1992). Foucault's oeuvre embodies precisely the tension that needs to be dealt with in order to develop an a-modern ethics of technology. While Foucault's early work focuses on the forces and structures that produce specific subjects, his later, ethical, work investigates how, amid these structures of power, human beings can constitute themselves as moral subjects. Humans are not only the 'objects' of power here, but also subjects that create their own existence against the background of and in interaction with these powers.

As Steven Dorrestijn (2006) has shown, this shift makes his work highly important for the ethics of technology. Foucault's ethical work makes it possible to develop an ethical framework that focuses on good ways of 'living with technology': deliberately shaping one's technologically mediated existence. In what follows, therefore, I will investigate to what extent Foucault's analysis of subject-constitution and his association with classical Greek ethics could form the basis for an ethical framework that can do justice to the technologically mediated character of human existence, and to the interwoven character of morality and technology.

The power of technology

Foucault is not generally considered a philosopher of technology. Only in the last period of his work, he explicitly used the word 'technology', using it to indicate what he called 'technologies of the self' – which are primarily existential techniques and not technological artefacts. Yet, his work can be read as a highly relevant contribution to the philosophy of technology. Foucault's work is generally divided in three periods, of which the first focuses primarily on knowledge, the second on power, and the third on ethics. Several authors have argued that especially his work on power is directly relevant to the philosophy of technology (Gerrie 2003, Sawicki 2003).

For Foucault, power is that which structures society and culture. Human existence does not take place in a vacuum, but in a world which is made of ideas, artefacts, institutions, organizations, et cetera, that all have an impact on human subjectivity. Vocabularies and scientific theories help to shape how we think; and our material environment organizes our actions, and social institutions and arrangements like schools, hospitals, armies and prisons give shape to how we live our lives, deal with illness, criminality, and madness. Technology, indeed, can be seen as one of these sources of power that help to shape the subject.

These powers, to be sure, for Foucault are not simply oppressive and alienating. Rather, they are *productive* in the sense that they produce the subject in ever new ways. And this production of subjects occurs in a very material and concrete fashion. Training practices for writing at schools, specific architectural designs of prisons, drilling techniques to learn to operate weapons, observation and surveil-lance techniques – all of these arrangements, that operate directly on the body, help a specific subject to come about. Subjects are produced by being 'subjected'. Specific forms of power introduce and enforce specific forms of normalcy and abnormalcy, which generate specific subjects.

Resistance and freedom

This analysis can be read in two different ways. A first reading focuses on the oppressive aspects of power. As Ladelle McWorther put it: 'If it is the case that power is the source of conscience and self-knowledge, then it would appear that individual selves have no control over their own beliefs and hence their own actions; agency is an illusion' (McWorther 2003: 114). This conclusion would bring us back to the conflict between the autonomy of technology and of human beings, with which I started this chapter. The only option we have in this reading is subversive action, in an attempt to liberate the subject from technological domination.

A second, and more adequate reading focuses on the productive aspects of power. As McWorther rightly showed as well, power should not be seen as external forces that operate on the subject 'from the outside'; rather, what subjects are and do, and how they understand themselves comes about in relations and networks of power. 'Selves are not constrained by powers external and foreign to them. Relations and networks of power *are* selves, *are* subjects' (McWorther 2003: 114). Rather than looking for a form of freedom that exists *outside* technology, this reading of power allows for articulating a concept of freedom in *relation* to power. In Jana Sawicki's words: '[F]reedom lies not in the discovery of essential features of the human situation, in complete mastery of reality, or in releasement'; it rather lies in the relations people develop toward the 'dominating powers of technology' (Sawicki 2003: 69).

Yet, it is crucial how one approaches such relations toward power. Sawicki, quoting Rajchman, describes the freedom of the subject in terms of 'rebelling against the ways in which we are already defined, categorized, and classified' (Sawicki 2003: 69). In this view, the freedom of the subject is to be found in resistance and opposition. But when power is what makes us the subjects we are, a merely subversive and rebellious attitude toward power does not offer a real alternative. Therefore I propose to articulate a more positive relation toward power – a relation that does not merely *oppose* power, but recognizes that subjectivity gets shape in interaction with power and that does not let this happen passively but actively *engages* in it. Rather than only undermining structures of power, the subject then takes them as a starting point, in order to contribute actively to the way in which they are constituted as a subject. Instead of being merely subversive and antithetic, the subject can also be engaged, trying to shape itself in the context of 'the powers that be'.

This relation to power opens up a different form of freedom. It is not to be found in the absence of influences that constrain the subject, but rather in *dealing with* these influences. Freedom becomes an *activity*; a practice of dealing with power, not a desirable final state of the subject in the absence of power. As Leslie Paul Thiele has argued, 'Foucault insists that freedom is not something to be secured, like the individual rights and opportunities that Isaiah Berlin described as negative liberty. Freedom is an activity to be engaged' (Thiele 2003: 225–26). In Foucault's own words: '(. . .) [T]he claim that "you see power everywhere, thus there is no room for freedom" seems to me absolutely inadequate. The idea that power is a system of domination that controls everything and leaves no room for freedom cannot be attributed to me' (Foucault 1997: 293). The freedom of the subject does not consist in being liberated from power, but in interacting with it. One becomes a subject not by securing a place outside the reach of power, but by shaping one's subjectivity in a critical relation to it.

This freedom-in-relation-to-power can also form the basis for a specific way of *dealing with* power. Freedom does not need to take the form of subversion and looking for an escape. In the words of James Bernauer and Michael Mahon: 'If one side to (. . .) resistance is to "refuse what we are", the other side is to invent, not to discover, who we are by promoting "new forms of subjectivity"'(Bernauer and Mahon 2005: 155). Dealing with power in practices of freedom opens the possibility to modify its impact on human subjectivity.

Ethics as governance

It is precisely this approach to freedom as a *practice of subject constitution* that connects Foucault's work on power and his later work on ethics. The late work of Michel Foucault opens a perspective on ethics which can serve as a basis to do justice to the intricate relations between ethics and technology, and to the technologically mediated character of moral action. In the last two volumes of his *History of Sexuality* he elaborates an ethical approach which differs radically from predominant ethical frameworks (Foucault 1990, 1992). For Foucault, ethics is not primarily about the question of which imperatives we need to follow and how we need to act, but about the question of how human beings constitute themselves as 'subjects' of a moral code. People do so in a productive relation to power, rather than in subversive attempts to undo or escape the effects of power – and this makes possible an interesting connection to the fundamentally technologically mediated character of human existence.

Foucault argues that any moral system or approach consists of three elements. Morality does not only encompass a *moral code* people have to comply with, and the *behaviour* corresponding to this code, but also a specific way in which human beings constitute themselves as *moral subjects* that follow this code (Foucault 1992: 25–32). A moral code of chastity, for instance, regulates the sexual behaviour of human beings and for doing so it requires moral subjects that organize their lives in such a way that they can subordinate their passions to the code. Ethics, for Foucault, primarily concerns this third element of morality: the ways in which human beings constitute themselves as moral subjects. The word 'subject' perfectly brings to expression that ethics is not only a matter of a person who is the 'subject' of his or her actions – like the grammatical 'subject' of a sentence – but that this person also 'subjects' him- or herself to a moral code; a specific vision of what constitutes a good life or good behaviour. The moral subject is not an *autonomous* subject; rather, it is the outcome of active *subjection*.

The moral subject has already taken many forms, like the Kantian subject that aims to keep its intentions pure, assessing them in terms of their potential to function as universal laws, or the utilitarian subject that aims to examine the consequences of its actions in order to attain a prevalence of positive outcomes over negative outcomes. These forms of 'subjection' or 'subjectivation' required by specific moral systems usually remain implicit. Ever since the Enlightenment has been self-evident to consider one's intentions and one's capacity to balance desirable and undesirable consequences as the proper ethical points of application. Foucault's approach to ethics questions the naturalness of this assumption. He argues that any form of ethics requires a specific moral subject, and is therefore necessarily based on a specific form of 'subjection'. Every moral system does not only define a code of behaviour but also a subject that is supposed to follow this code. Also following the Kantian categorical imperative or the consequentialist principle of utility requires a specific moral subject, after all, which 'subjects' a specific aspect of one's person to specific criteria.

In classical Antiquity, however, Foucault discovers an ethical approach that did not *implicitly* define a specific moral subject, but *explicitly* directed itself at the constitution of one's moral subjectivity. Foucault's investigations regarding classical ethics were primarily directed at sexuality. He argues that in ancient Greece, sexuality was not primarily organized via a moral code of imperatives and prohibitions, but rather in terms of *styling* one's dealing with pleasure. Ethics consisted in finding such a relationship to one's sexual desires and drives that these do not *determine* the self but become the object of active 'design'. This 'design' or 'styling' took place in the form of what Foucault called 'self practices'. Such practices were ways to experiment with and give shape to the one's way of dealing with pleasure. Foucault also indicated them as 'technologies of the self' (Foucault 1997: 223–52).

Rather than simply following one's passions and desires, self practices aimed at gaining a productive distance. From this distance, not pleasure but the subject itself can have the central place in determining how one lives life. In a variety of ascetic and aesthetic practices, the subject got shape in an explicit way, rather than being a side product of following a specific code. Ethics was a matter of 'care of the self': paying careful attention to one's subjectivity, and shaping one's life in a desirable way.

Ethics of mediated subjectivity

What can this ethics of self-constitution imply for the ethics of technology? Foucault's ethical perspective unites two elements that usually remain opposites in ethics: the radically mediated character of human actions on the one hand, which causes the subject to lose the autonomy it used to have ever since the Enlightenment; and the ability of the subject to relate itself to the powers that help to shape it on the other hand, which makes it possible to modify their impact. While Foucault directed his attention to the role of the pleasures in ancient Greece, in our technological culture technology is a pre-eminent example of the powers that help to shape the subject. Because of their fundamentally mediating role in people's actions and interpretations, technologies implicitly shape what constitutes a good life. By finding an explicit relation to these technological mediations, incorporating them deliberately in our existence rather than subscribing unknowingly to their impact, human beings can take explicit responsibility for their technologically mediated existence.

Autonomic computing plays a particularly challenging role here. By designing technologies that interfere deliberately and intelligently but often unnoticed with our intentions and our behaviour, we have started to trust part or our morality to technology. Think of technologies like the *HygieneGuard*, which is a system for children's toilets that reminds kids to wash their hands after using the toilet in case they forget to do so, and the *FoodPhone*, which is a specific application of mobile telephones with built-in cameras to help obese people lose weight, requiring you to take a photograph of everything you eat and to send it to a central office, after which you receive feedback on the number of calories you have had. To such technologies we explicitly delegate aspects of our morality, trusting that they will educate us in desirable ways.

The same holds true for intelligent cameras that can recognize faces, or that can discriminate between regular and aggressive behaviour. To these cameras we trust forms of supervision that were traditionally done by human beings – including the risks of failure or implicit bias. Automatic face recognition systems, for instance, sometimes abusively identify a person as suspect – which actually appears to happen more often for people with a dark skin and for older people, because the software in these systems is tuned to light contrast on a young, white skin (cf. Introna 2005: 75–86). How to find good ways of 'living with technology' in these cases?

A first answer to this question is that we should move beyond the predominant ethical question if specific technologies should be allowed or not. In many cases, they are there already, urging us to find ways to embed them in society in responsible and desirable ways. Just like the Greeks in classical antiquity did not have to deny or renounce the pleasures to be a moral subject, neither do we have to deny the technologically mediated character of our existence to be moral subjects. If technology fundamentally mediates what kind of humans we are, this does not yet imply that 'humanity' is mastered by 'technology' or that 'the system' has entered 'the lifeworld' and causes humans not to be treated as subjects but as objects, as the Heideggerian and Habermasian positions want us to believe. From a Foucauldian perspective, the technologically mediated character of life in a technological culture does not necessarily pose a threat to the subject. As Steven

Dorrestijn has elaborated in his study on the ethics of behaviour-influencing technology, technologies rather form specific ways in which the subject is constituted, which can be the starting point for moral self practices (cf. Dorrestijn 2004: 89–104).

Ethics, then, should not have the character of protecting 'humanity' against 'technology', but should consist in carefully assessing and experimenting with technological mediations, in order to explicitly shape the way in which they help to shape subjects in our technological culture. In our contemporary technological culture, self practices consist of deliberately using and designing technology, anticipating and modifying its mediating role in our existence, and realizing that each way of using it also helps to shape one's subjectivity. Such self practices could be described as 'techniques of using technology'. They require a form of ascesis, without implying, again, that one should refrain from technology, or only use it reluctantly from a Heideggerian attitude of 'releasement' (*Gelassenheit*). Technological ascesis rather consists of using technology in a deliberate and responsible way, such that the subject that emerges from it – including its relations to other people – acquires a desirable shape.

Gaining a free relation to technology

For Foucault, the *telos* of subject constitution is freedom. Again, this notion of freedom does not consist in an absence of power, but in gaining a specific relation to it. As Foucault said in an interview: 'I am sometimes asked: "But if power is everywhere, there is no freedom." I answer that if there are relations of power in every social field, this is because there is freedom everywhere' (Foucault 1997: 292). Not power should be rejected, but domination, as the 'perversion of power' that takes away the possibility of freedom (O'Leary 2002: 158).

Freedom in Foucault's work thus functions both as the condition for ethics and as its ultimate aim (O'Leary 2002: 154–70). On the one hand, there can only be ethics when people are not completely dominated by power; on the other hand, ethics consists of developing 'practices of freedom' in which people interact with power to constitute their subjectivity. As such, the Foucauldian concept of freedom offers an interesting alternative to the criterion of autonomy that is often used in ethical theory. While the concept of autonomy stresses the importance of the absence of 'external influences' in order to keep the moral subject as pure as possible, the concept of freedom recognizes that the subject gets shape in *interaction* with these influences. The subject is not what remains when all powers and mediations are *stripped* from it; it is what results from an active design and styling of the impact of these powers and mediations. The core of a Foucauldian ethics of technology is: gaining a free relation to technology, which allows one to style the way in which one's technologically mediated subjectivity gets shape.

This implies that not all technological innovations should be embraced as desirable. A first requirement for such self practices to come about, after all, is the very possibility to develop a free relation to technology. Such a free relation – not to be confused with autonomy – can only exist when the power exerted by technological mediation does not turn into complete domination. In such cases, people's subjectivity cannot be the outcome of a subtle interplay between technological mediation and human appropriation anymore; human beings are entirely determined by technology then.

It must be noted, though, that in practice many technological innovations do indeed offer the possibility to develop a free relation to them, even though they might seem oppressive at first glance. Technological possibilities in anaesthesia, for instance, as Gerard de Vries elaborated (De Vries 1999), were initially seen as a threat to humanity, but by now they have changed what we consider to be human dignity. Operating on somebody *without* anaesthesia has come to be something immoral by now.

We must be very careful, therefore, when drawing a boundary between mediating power and oppressive domination. Every technology puts at stake what it means to be a human being, and only when the relations between human beings and new technologies are starting to crystallize we can develop an explicit relation to their role in our existence. Using technology is a learning process, in which technologies become domesticated and in which human beings are mediated in new ways. The social role of the cell phone is a good instance of this. While initially it was socially not really accepted to have a phonecall in public, after some time people started to use the cell phone almost unlimitedly. By now, codes for cell phone use have developed, like minimizing the duration of mobile calls in public transport; turning off the ringsound during lectures, concerts, or meetings; and a deliberate use of the voicemail and texting functionality as alternative forms of contact. In Dutch trains, after a ban on smoking, former smoking compartments now even function as 'silence compartments', where cell phone use and conversations with other passengers are forbidden.

In the field of autonomic computing, the borderline between mediating power and dominating force is very subtle as well. One of the key issues addressed above, after all, is the very impact of such technologies on human freedom. When smart environments take over control over part of our actions, this can easily be seen as a form of domination. But this cannot be the right conclusion in all cases. As expressed in the *Phaedrus* dialogue, Socrates was against the technology of writing, because it would destroy our memory and weaken our minds. But fortunately, this oppressive form of domination turned out to be a very interesting reconfiguration of humanity. Without writing, indeed, we would not have been who we are now. But what initially seemed to be oppressive turned out to be constitutive for a new dimension of being human. Likewise, we will have to reinvent what it means to be a free, conscious, and responsible person in our interactions with smart environments and persuasive technologies.

This does not happen automatically, though. It requires that citizens of a technological society be equipped with a critical attitude toward technology, which allows them to see through their functionality and recognize their mediating roles. Moreover, *using* technologies in deliberate ways is not the only way to modify and stylize technology's impact on human subjectivity. Also the *design* of technologies can be an important 'self practice' in our technological culture. Any technological design is the starting point of an artefact that is not merely instrumental but that also helps to shape the subjectivity of its users. As I elaborated elsewhere (Verbeek 2006, 2008b), designing technologies explicitly from this perspective can be both an enrichment of existing design methods, and an important way to 'care for the self'.

This means that, rather than merely opposing smart mirrors, automatic speed influencing, and face recognition systems, we need to develop ways of shaping our lives in interaction with them. By understanding the hidden normativity in smart mirrors, for instance, which enables people to see the feedback given by these devices as an interesting yet biased perspective. Or by learning to trust ourselves to a system that automatically limits the speed of our cars, which can enable us to experience this not so much as an infringement of human freedom but as a different way to organize and condition transportation, which also has important benefits (cf. Dorrestijn 2004). Even automatic face recognition systems can then be seen as part of an ongoing experiment to reorganize the boundary between public and private life – just like the data of our cell phones show where we are, and our profiles on social networking sites show what we have done. This is not to say that all of these technologies and their applications are desirable. What these examples show is that we should not be fixated on the question if these technologies should be allowed, but rather on the question of what could be good ways to give them a place in society.

Conclusion: moral agents and mediated subjects

Autonomous computing opens a new configuration of humans and technologies, that does not fit the classical configuration of 'using' technology anymore. Technologies start to form interactive and intelligent environments in which human existence plays itself out, while their mediating role in human actions and experiences remains often unnoticed. However tempting it may be to classify this new category of technology as a threat to human freedom, privacy, and dignity, though, a closer inspection of these 'smart environments' shows that they rather radicalize the phenomenon of mediation that has always accompanied technological developments. Human beings cannot be understood without taking into account how technology shapes their existence – and autonomic computing adds an interestingly new dimension to the technologically mediated human condition.

This does not imply, however, that we should give up on ethics of technology. Rather, we need to move beyond the predominantly externalist approach in the ethics of technology. Instead of an ethics of 'technology assessment' that places itself outside technology we need an ethics of 'technology accompaniment' that recognizes how deeply humanity and technology are intertwined. Ethics should not primarily direct its attention at the question if specific applications of autonomic computing are morally right or wrong, but at the question of how to live with these technologies.

The ethical work of Michel Foucault appears to provide an interesting framework for just such an 'ethics of the technologically mediated good life'. It opens an ethical perspective that is centred around the question of how to shape one's subjectivity in a context of power relations. In the context of technology, this ethical approach requires us to develop a free relation toward technology, in which we recognize that every technology inevitably mediates human existence, which explicitly helps us to shape the ways in which these mediations affect our subjectivity.

An ethical approach like this enables human beings to take responsibility for technology. In fact, it is the very refusal to take technologies seriously which marginalizes an ethics of technology from the outset. After all, the technological developments themselves continue to move on, and while squeaky-clean ethicists grumble on the sidelines, they are missing the opportunity of contributing towards the responsible development and the responsible use of autonomic computing. The world is already full of intelligent technologies, and smart environments will become ever more widespread. It is high time that ethics moved on from considering simply whether or not these are acceptable and started addressing the issue of the best way to embed such technologies in our society.

Notes

1   This article incorporates reworked fragments from my inaugural lecture 'De grens van de mens: over techniek, ethiek en de menselijke natuur' (Enschede: University of Twente, 2009) and from my book *Moralizing Technology: Understanding and Designing the Morality of Things* (Chicago: University of Chicago Press, forthcoming 2011). This research was made possible by the Netherlands Organization for Scientific Research (NWO), as part of the NWO-VIDI project 'Technology and the Limits of Humanity'.

References

Aarts, E. and Marzano, S. (2003) *The New Everyday. Views on Ambient Intelligence*, Rotterdam: 010 Publishers.

Bernauer, J. and Mahon, M. (2005) 'Michel Foucault's Ethical Imagination', in Gary Gutting (ed.) *The Cambridge Companion to Foucault*, Cambridge: Cambridge University Press.

Dorrestijn, S. (2004) *Bestaanskunst in de technologische cultuur: over de ethiek van door techniek beïnvloed gedrag*. Master's thesis, Enschede: University of Twente.

—— (2006) *Michel Foucault et l'éthique des techniques: Le cas de la RFID*, Nanterre: Université Paris X (Mémoire).

Ellul, J. (1964) *The Technological Society*, New York: Vintage Books.

Fogg, B.J. (2003) *Persuasive Technology: Using Computers to Change What We Think and Do*, San Francisco: Morgan Kaufmann/Elsevier.

Foucault, M. (1990) *The Care of the Self – The History of Sexuality, vol. 3*, London: Penguin Books {1984}.

—— (1992) *The Use of Pleasure – The History of Sexuality, vol. 2,* London: Penguin Books {1984}.

—— (1997) *Ethics: Subjectivity and Truth* (ed. Paul Rabinow), New York: The New Press.

Freud, S. (1989) *Inleiding tot de Psychoanalyse*, Meppel: Boom {1917}.

Gerrie, J. (2003) 'Was Foucault a Philosopher of Technology?', *Techné: Research in Philosophy and Technology* (7) 2: 14–26.

Habermas, J. (2003) *The Future of Human Nature*, Cambridge: Polity Press.

Ihde, D. (1990) *Technology and the Lifeworld*, Bloomington/Minneapolis: Indiana University Press.

Introna, L. (2005) 'Disclosive Ethics and Information Technology: Disclosing Facial Recognition Systems', *Ethics and Information Technology* 2005–7: 75–86.

Latour, B. (1991) *Nous n'avons jamais été modernes*, Paris: La Découverte.

McWorther, L. (2003) 'Subjecting Dasein', in Alan Milchman and Alan Rosenberg (eds.) *Foucault and Heidegger: Critical Encounters,* Minneapolis: University of Minnesota Press: 110–26.

O'Leary, T. (2002) *Foucault: The Art of Ethics*, London/New York: Continuum.

Sawicki, J. (2003) 'Heidegger and Foucault: Escaping Technological Nihilism', in Alan Milchman and Alan Rosenberg (eds.) *Foucault and Heidegger: Critical Encounters*, Minneapolis: University of Minnesota Press: 55–73.

Sloterdijk, P. (1999) *Regeln für den Menschenpark: Ein Antwortschreiben zu Heideggers Brief über den Humanismus,* Frankfurt/M: Suhrkamp.

Thiele, L. P. (2003) 'The Ethics and Politics of Narrative: Heidegger + Foucault', in Alan Milchman and Alan Rosenberg (eds.) *Foucault and Heidegger: Critical Encounters*, Minneapolis: University of Minnesota Press: 206–34.

Verbeek, P.P. (2006) 'Materializing Morality – Design Ethics and Technological Mediation', *Science, Technology and Human Values* (31) 3: 361–80.

——— (2008a) 'Cyborg Intentionality – Rethinking the Phenomenology of Human-Technology Relations', *Phenomenology and the Cognitive Sciences* (7) 3: 387–95.

——— (2008b) 'Morality in Design: Design Ethics and the Morality of Technological Artifacts', in Pieter E. Vermaas, Peter Kroes, Andrew Light, Steven A. Moore (eds.) *Philosophy and Design: from Engineering to Architecture*, Dordrecht: Springer: 91–103.

——— (2008c) 'Obstetric Ultrasound and the Technological Mediation of Morality – A Postphenomenological Analysis', *Human Studies* (31) 1: 11–26

——— (2009) 'Ambient Intelligence and Persuasive Technology: The Blurring Boundaries between Human and Technology', *Nanoethics* (3) 3: 231–42.

Verbeek, P. P. (forthcoming 2011) *Moralizing Technology: Understanding and Designing the Morality of Things*, Chicago: University of Chicago Press.

Vries, G. de (1999) *Zeppelins – over filosofie, technologie en cultuur,* Amsterdam: Van Gennep.

Chapter 3

Remote control

Human autonomy in the age of computer-mediated agency

Jos de Mul and Bibi van den Berg

'Je est un autre'

(Rimbaud 1871)

Introduction

Human beings have always used instruments, media and machines to strengthen and expand their agency. These technologies enable them to have 'remote control' over both the natural and human world. Technological extensions serve to increase the 'action radius' of human autonomy. They enable us to do things we couldn't do without them: writing makes it possible for us to delegate our memories to clay tablets, papyrus or paper. Pulleys facilitate lifting things that are far too heavy for our human bodily constitution. Telephones and e-mail enable us to be socially present in places while being physically absent from them (cf. Gergen 2002). Gamma knives allow us to target brain tumors with high doses of radiation therapy without affecting (much of) the surrounding tissue. And the Mars Exploration Rover enables us to gain insight into the geological history of Mars under circumstances that are physically impossible to survive for humans.

However, as the human life world transformed from a 'biotope' into a 'technotope' in modern culture, a fear emerged that human beings would become dependent on, or even slaves of technology (cf. Ellul 1988, Heidegger 1962). This dystopian perspective of the technological world is all the more worrying, to its adherents' mind, because the *responsibility* for that world and what happens in it is still in the hands of human beings and not in the hands of the technologies. After all, human beings are the architects, designers and users of technologies, and for that reason they are responsible for their creations and their creations' output.

With the advent of 'autonomic computing' – ubiquitous computing, Ambient Intelligence, pervasive computing, expert systems, artificial intelligence, artificial life, converging technologies, etc. – it seems that we can no longer understand these matters in a merely metaphorical sense. Autonomic computing appears to mark the transition into a phase in which technologies actually gain *agency* and become a potential threat to human autonomy.

In this chapter we will argue that this fear is excessive, because it starts from a misleading opposition of human agency and technical artefacts. Discussing the intimate relationship of man and technology, we will develop a notion of autonomy that focuses on the concept of '*remote control*'. We will argue that autonomic computing does not necessarily form a threat to our agency, but that, quite to the contrary, it may *strengthen* it. Note that we do not claim that autonomic computing *necessarily* strengthens human agency and autonomy. The most pressing question, we argue, is not whether autonomic computing strengthens human agency or not, but rather *under which circumstances* it does, and under which circumstances it threatens human agency. We will investigate this question by discussing a number of real and fictional cases dealing with increasingly more radical instances of 'autonomic voting'.

Electoral compass(ion)

Perhaps one of the social phenomena in which we express our human autonomy most explicitly is that of democratic elections. In elections our choices, made freely and on the basis of (rational) arguments, may contribute to the maintenance and management of our society. During elections we must use (explicit) reasoning to choose which political programme we approve of most, which concrete policies we endorse, and which political ideals we would like to see realized.

As is the case in many Western countries, in the Netherlands this is no easy feat: there are numerous political parties and whoever makes it his explicit goal to choose responsibly, must have access to the right information (both in terms of channels and in terms of content) with regard to the political agendas of all these parties. Thankfully, there are many ways to go about getting this information: electoral meetings, paper and online party programmes, flyers, websites, commercials on television and message boards on the street, election debates on television, etc. However, despite all these sources of information, research shows that there are relatively few Dutch citizens who inform themselves elaborately and take the time to study all party programmes.[1] Of course many reasons can be cited for this behaviour, ranging from lack of interest to laziness, and from feeling politically underrepresented to downright rejection of the democratic system as a whole. Another reason, which is less often cited, may be information overload. Who has the time to read all of these party programmes, to watch every election debate, to go to every political rally? This is the reason why many voters vote based on their gut feelings, or don't vote at all.

This phenomenon was clearly visible in the democratic elections held in the Netherlands for the 'Water Boards' in the fall of 2008. In a country of which an important part of its landmass is below sea level the Water Boards are *literally* of vital importance. The Boards' task is to maintain dams and dykes and protect the country from both seawater and river flooding. The Water Boards are one of the oldest institutions in the democratic system of governance in the Netherlands – the first Water Board was created in 1122 AD, so the Boards have a long history indeed. The Water Boards literally form the backbone of the Dutch 'polder model': traditionally each Board regulates the water maintenance for a specific region of the country. This means that their main responsibility is to guarantee three goals: flood defenses, preserving water quality, and managing the general water economy of the region.

Until recently the Water Boards focused predominantly on technical management. Lately they have politicized to some extent. As a political institution the Boards used to consist of *individual* elects, who could be chosen as members of the Board for a four-year term. In the elections of November 2008 political parties were introduced for the first time in all 26 Boards nationwide. Partially, these consisted of the traditional political parties that also populate the Dutch parliament, such as the Christian democrats (CDA), the (neo)liberals (VVD) and the labor party (PvdA). But there were also two new national parties, which can only be elected in this specific election: Water Natural ('Water Natuurlijk') and the General Water Board Party ('Algemene Waterschapspartij'). The central themes in this election were the projected rising sea level as a result of global warming on the one hand, and environmental conservation on the other hand – two themes that do not necessarily go well together, and that even stand in serious opposition in many cases.

Despite the fact that the political importance of an institution such as the Water Boards has increased considerably in light of both of these themes, the number of voters for this election turned out to be dramatically low. Only 24 per cent of the adult population voted (of which, notably, a marked number turned out to be unlawful, because the voting bill was too complex and voters did not fill out the form correctly). The authors of this chapter also struggled with the question of what to vote, but both felt it was their democratic duty to take part in the elections. They turned to ICTs for help on the matter.

Kieskompas enterprise, a Dutch private enterprise in which entrepreneurs and social scientists from the VU University of Amsterdam collaborate, designed a unique Electoral Compass for every Water Board. The Compass consisted of 36 theses in 12 categories: taxes, governmental innovation, democracy, dykes and roads, economy and environment, energy and climate change, nature and recreation, plants and animals, polders and landscape, water, floods, and living and working. These theses had to be valued on a five-point scale (completely agree, tend to agree, neutral, tend to disagree, completely disagree – alternatively, users could choose 'no opinion'). After completing the questionnaire, one could see one's position in the political landscape as it aligned with the parties participating in one's own Water Board, and thus find out which party would represent one's preferences best. The political landscape was represented along two axes, one ranging from ecology to economy, the other from a broad to a small range of responsibilities for the Water Boards. The Electoral Compass also provided information about the history and tasks of the Water Boards, and voters were given information regarding the policy proposals of the participating parties. It turned out to be very popular, not only in this election but in any of the democratic elections for which it was developed (elections for the national parliament, local elections, and even the US presidential election). It did, in fact, help us to make a choice for a party in the Water Board elections.

What is interesting about the case of the Electoral Compass as a technology is this: in its current form we delegate the task of delving through all the information involved in a single election to the technology, but the individual voter remains in control of the final casting of the vote. The technology sifts through all the information to provide us with an easier choice, which means that it has taken over the *process* of collecting and weighing all the information, yet the voter himself makes the final decision – the goal of voting for an election is still in the hands of the human agent.

The fact that the Electoral Compass leaves the final decision up to the voter is one of the main differences with another version of this type of technology in the Netherlands, called VoteMatch (StemWijzer). The latter, which has been developed by the Dutch Centre for Political Participation (Instituut voor Publiek en Politiek, IPP), gives a clear voting advice, whereas the Electoral Compass does not. The Electoral Compass only aims at providing the user with an easy overview of his personal alignment in relation to various parties in one election. Another difference is that VoteMatch only offers three options to the theses presented (agree, disagree, don't know). On the other hand, VoteMatch provides the possibility to indicate the relative weight of each topic whereas the Electoral Compass does not. Yet another voting aid, the ChoiceAdviser (KiesWijzer), does not present theses, but questions to the user. There are only ten of them, and the user has to choose between three different answers (and 'no opinion'). After answering the questions, the user is shown the amount of affinity he has with each of the political parties, and with a table with the answers of all parties to these questions.

After the introduction of these technological voting aides each of them received critiques regarding the type of questions they pose, their objectivity and transparency, the way they present results, and the type of advice they provide. The Electoral Compass has been accused of being one-sided and even strongly biased, because it favours parties in the centre of the political landscape. We will come back to this below. What became clear is that technological voting aides are *never fully neutral*, but always reflect, to a certain extent, the technological, methodological and political decisions that have been made by the designers of these aides.

Such biases have been researched extensively in relation to various technologies under the name of 'scripts'. Three different meanings of the term script can be distinguished in relation to technologies (Van den Berg 2009). First, there is the idea that designers implement ideas about users and prospected contexts of use into the technologies they design. For instance, conceptions of users in terms of gender may result in product designs that are not only very different for men and women, but also *express* this difference in their materiality. Van Oost has shown that shavers for women and for men are different in shape, in the buttons they have, and in the ways they can and cannot be used. She concludes that the designers and manufacturers of these products not only sell shavers for men and women, but also affirm and reify gender (Van Oost 2003: 207). This line of research has come to be known as 'script analysis' and over the years has come to play an important role in Science & Technology Studies (cf. Akrich 1992, 1995, Berg 1999, Gjøen and Hård 2002, Latour 1992, Van Oost 2003).

Second, in artificial intelligence the term script refers to the human ability to quickly and easily come to understand a wide array of everyday, recurring and ritualistic 'scenes' and to know how to act in them (Schank and Abelson 1977). For instance, when entering a restaurant, we instantly know that a cycle of actions such as 'finding a seat', 'reading the menu', 'ordering food', 'eating the food' and so on can be expected. Human agents apparently possess a type of knowledge, called scripts, to deal with such standardized, regularly occurring sequences of action related to specific scenes. Artificial intelligence research aims at making such scripts explicit, so that they can be mimicked by computer technologies, which as a result, the argument goes, will lead to smarter, more life-like machines.

Third, artefacts *themselves* may act as scripts, sometimes as intended by their designers, and sometimes unintentionally. This is evident for instance in the way in which a groom on a door, that makes the door harder to open, discriminates against specific groups of users, most notably older people, children, and people carrying things in their hands (Latour 1992).

Technological artefacts have the interesting characteristic that they often *contain* scripts in the first two senses discussed here, but *are* scripts themselves (the third sense) as well (Van den Berg 2009). In the example of voting aides discussed here this is evident. First, as we have seen one of the critiques against voting aides is that they contain the designers' ideas – both political opinions and conceptions of users – that may influence the opinions and voting behaviours of users. For instance, the Electoral Compass, developed by the VU University in Amsterdam, a protestant university, we have seen above, was accused of favouring parties in the political centre, and most notably the Christian democrats (CDA). The designers' political preferences are thus scripted into the Compass's design. These scripts embedded into the software may (implicitly) steer the voters' eventual choices. This is the meaning of scripts in Science & Technology Studies. Second, the voting aide uses ideas about the way in which people generally tend to choose between different bits of information given to them and how they value these various offerings. These ideas are translated into algorithms that thus represent scripts in the second meaning of the term, as presented by artificial intelligence. Third, the voting aide *itself* has a script force in the sense that it steers people in certain directions. When the user completes the answers to the questions, he will feel he has every reason to believe that the choice presented to him by the voting aide represents, or even *is* in fact, his own choice. Particularly this last point is interesting in light of developments in the direction of autonomic computing, and therefore we will come back to it more extensively below.

We may conclude that all voting aides ultimately raise the same question: do such technologies threaten or undermine our autonomy? Ought we not to decide for ourselves, as individual human agents, what party we want to vote for? After all, as such voting aides contain hosts of scripts and choices, they affect our behaviour and therefore guide us in directions that we may not have chosen, had we taken the time and put forth the effort to gather all the information required to vote responsibly for ourselves. In short, *voting aides seem to undermine our (political) autonomy.* Surprisingly, this effect on our autonomy applies to the very domain in which human beings claim to express her most strongly: the design and use of technological artefacts.

Autonomy and distributed agency

Our modern Western (liberal, secular, scientific) culture values human autonomy as one of the pillars – if not the most important pillar – of agency.[2] Some argue that this emphasis can be traced back to the philosophy of René Descartes (Gontier 2005). In Descartes' philosophy the rationality of human agency is not only underlined, but even *assumed* as the starting point from which we ought to understand the reasons that motivate human action. Over time this assumption in Descartes' work has been criticized with various arguments. One line of criticism points towards the fact that there are numerous *other* factors that motivate human agency apart from rationality, and that motivate human agency to such an extent that our rationality only has a fraction of the importance that we generally ascribe it. Human agency, some of these criticists say, is predominantly the result of our genetic makeup, or of our upbringing, as others claim, or of our social class, or our gender, as yet others argue (Rosenberg 2008). These lines of criticism all lead to harder or softer forms of determinism – human agency, in their view, is largely the result of forces that are either biologically or socially determined, or a mixture thereof, and that, in any case, fall outside the sphere of complete control of the individual agent. Contra Descartes it is generally held nowadays that human agency is the result of a *complex* of various factors, which includes both our rational deliberation, and a set of 'natural' forces such as our genes and passions, and of 'nurture' forces such as our upbringing, social class, gender, etc.

A second line of criticism confronting the Cartesian emphasis on the rationality of human agency starts from the factual claim that in the vast majority of our actions rational deliberations are at best implicit, but more often than not wholly absent. When rational deliberations *do* play a role in our actions, most of the time these deliberations emerge retrospectively – one formulates reasons for one's actions *post actio* rather than *pre actio*, or even *in actio*. Apart from a longstanding tradition in philosophy that debates the alleged freedom of choice that human agents have, research in neurosciences has empirically shown that there are many instances in which our brain have already 'made a decision'[3] before we even become *aware* of what we want to do (Libet 1985, Burns and Bechara 2007).

These two lines of critique have given rise to a postmodern denial of the existence of autonomy and free will, both in the natural sciences (cf. Dawkins 2006) and in the humanities (Nietzsche and his postmodern heirs). However, we argue that this extreme position takes matters too far. In our everyday lives we experiences ourselves *as agents* with *some level of control and freedom of choice*, even if we are willing to grant the postmodern suspicion that our levels of control are far from complete, and that our freedom of choice may often be informed by motives and processes that are either unknown to us or principally not (completely) insightful for us. Our *experience* of ourselves *as agents with some level of control* is ignored or denied by both the natural scientists and the postmoderns in the humanities. However, we feel that this first-person perspective needs to be taken seriously, as it seems real to us in everyday life, despite the limitations of its reality as pointed out by these scientists. Obviously, having *incomplete* control does not imply having *no* control *at all*.

As we have seen above, human agency is the result of a number of factors combined, including biological, social, and cultural ones, and including our rational facilities for deliberation. What we want to emphasize is the fact that, although we have limited control over the forces that motivate our action and the elements that become part of our identities, human agency has a distinctive ability to *affirm* its actions in a unique manner: via a '*reflexive loop*'. This reflexive loop enables us to view and judge the internal and external forces that motivate us *as forces that motivate us* to act, and to affirm and embrace those actions *as our own* (or, alternatively, to distance ourselves from them). One could argue that what happens in the reflexive loop is that we temporarily leave our first-person perspective and take a third-person, a *remote* perspective towards ourselves and our own actions. In this remote stance we gain a certain amount of freedom towards the forces that drive us.

The human constitution varies from that of plants and animals in the sense that a human being not only *is* a body (as is a plant), nor *is* and *has* a body (as does an animal), but *is*, *has*, and simultaneously can always *relate to* its body from an external position. Or, to phrase it in experiential terms: 'Man not only lives (*lebt*), and experiences his life (*erlebt*), but he also experiences this experience of life' (Plessner 1981: 364, also see De Mul 2003). This latter fact is precisely the reason why we are always engaged in a reflexive loop: we *can* view and judge our actions from a distance – though we need not always do so.

A classical example of the workings of this reflexive loop, and of autonomy in relation to actions, can be found in Euripides' tragedy *Medea* (Euripides 2006). In this Greek tragedy we encounter Medea, who has been left by her husband Jason for a younger woman. She is furious, feels utterly humiliated and therefore seeks revenge. Tormented by conflicting emotions she struggles to weigh up options for vengeance and in the end she chooses to kill their children to get back at Jason. *Medea* is often cited as the first example of the expression of free will. In fact, this interpretation is at best one-sided. Medea does not express free (Cartesian) agency – rather, she is motivated by various forces, and is ripped apart between clashing emotions: on the one hand the hatred she experiences with respect to her (ex)husband Jason, who has left her in the most humiliating and abominable way, and on the other hand the love of her children. Both of these emotions battle for dominance within the person of Medea. Medea, therefore, is not 'free' in the sense that she can make a decision based on pure rational deliberation. One could even argue quite the opposite, with the postmoderns above: that Medea does not have agency at all, because she is ruled by her passions (or 'daemons', in the language of Euripides).

However, this is not the case either. Medea *is* in fact an agent, because in the process of struggling with these clashing forces inside her, forces that pull her to this side and that, she makes a decision and affirms one force at the cost of another. She ends up embracing her hatred for Jason as the main motivator for her actions and thus decides to kill her children. While this horrible death may make her an unsympathetic character, we have to grant Medea the fact that she does take responsibility for her *daemon*. She embraces the action she completes. She identifies herself with her action, despite the fact that it originated in an overwhelming force over which she had little control. Therefore, Medea is in fact a good example of what one could call 'responsibility without freedom' (Alford 1992). And this, we argue, is precisely the minimal requirement of what it means to have human agency (cf. De Mul 2009: 179–244).

Human agency, we argue, is based on *reflexive remote control*: our actions are remote(ly) controlled, that is, they are motivated, stimulated, challenged, and shaped, by countless internal and external forces, but as reflexive beings we simultaneously exert remote (self-)control *via* the forces that motivate us. This is comparable to the zapper handling a remote control in front of the television: although he has a rather limited influence on the choice of television shows that are on TV, nevertheless handling the remote control enables him not only to make choices with regard to which shows and channels he wants to watch, but more importantly: he is responsible for his choices, for the self-reflexive cycles they engage, and therefore for the 'bricolaged' identities he zaps together.

What the example of Medea shows is that external and/or internal forces outside our deliberative faculties may limit our human autonomy. In extreme cases, such as blind anger or senseless panic this is indeed the case. In many other cases, however, we find forms of externalization that don't undermine, but rather enhance human autonomy instead. Dilthey has shown that, contrary to the rationalistic, introspective tradition instigated by Descartes, our *lived experiences* (*Erlebnisse*) more often than not are far from transparent. Our thoughts, motives and feelings often remain implicit and we only get to know them or gain insight into them in the process of *expressing* them (*Ausdruck*), that is: in speaking, in the language we use, in our actions, in the clothes we wear, in the laws we write and adhere to, in the institutions we construct and embrace, etc. Implicit meanings, ideas and feelings are articulated in our expressions, and thus instigate an *understanding* (*Verstehen*) of ourselves, of our motives, and our drives. Dilthey explains this autonomy-enhancing reflexive loop as follows:

An expression of lived experience can contain more of the nexus of psychic life than any introspection can catch sight of. [. . .] In lived experience we grasp the self neither in the form of its full course nor in the depths of what it encompasses. For the scope of conscious life rises like a small island from inaccessible depths. But an expression can tap these very depths. It is creative.

[Finally] it is the process of *understanding* through which life obtains clarity about itself in its depths [. . .] At every point it is understanding that opens up a world.

(Dilthey 1914–2005: 206, 220, 87, 205, also see De Mul 2004: 225–56)

Technologically mediated agency

In the evolution of the human life form cognitive artefacts have played a crucial role in the reflexive loop of lived experience, expression, and understanding. The act of writing is a good example. Since the so-called 'mediatic turn' in the humanities – initiated by McLuhan and his Toronto school – much attention has been paid to the fact that writing is not just a neutral instrument to express thoughts, but structures and enhances human thought in specific ways. In his book *Orality and Literacy* Walter Ong has argued that the transformation of oral cultures into writing cultures opened a whole new domain of human agency and culture:

> Without writing, words as such have no conceivable meaning, even when the objects they represent are visual. They are sounds. You might 'call' them back – 'recall' them. But there is nowhere to 'look' for them. They have no focus and no trace (a visual metaphor, showing dependency on writing), not even a trajectory. They are occurrences, events. [. . .] By separating the knower from the known, writing makes possible increasingly articulate introspectivity, opening the psyche as never before not only to the external objective world quite distinct from itself but also to the interior self against whom the objective world is set.
>
> (Ong 1982: 31, 105)

Without the use of these very fundamental artefacts, which me may call 'external devices of reflection', humans as we know them would not exist. The reverse, of course, is true as well: it is human beings that create and develop artefacts, and that interpret them *as artefacts* in their use, hence without them these artefacts would not exist, neither practically nor ontologically.

Interestingly, initially writing met severe criticism, because it was not so much understood as an autonomy-enhancing technology, but rather as a threat to human autonomy. For example, in *Phaedrus* Plato critically discusses the invention of writing and what he conceives to be the downfall of both oral culture and human memory (Plato 1914, Phaedrus 275A). In this dialogue the Egyptian king Thamus argues that writing will eliminate the human capacity to remember, because humans will forget to practise their memories. By delegating human abilities to technological artefacts, Thamus reasons, humans will lose powers, capabilities, sources of agency. The underlying theme voiced by Plato in this dialogue is a perspective regarding human *autonomy* in relation to technological artefacts – one that has been echoed many times over in recent decades with regard to the delegation of cognitive tasks to computer technologies. In these modern variants of Plato's argument the central line of reasoning is not that the *products* of our thinking are delegated to technological artefacts, as was the case in *Phaedrus* discussion on writing and memory, but even worse: important parts of the rational and moral *process* of thinking itself is delegated to computers. According to these critics, this will lead to an undermining of human autonomy.

However, Plato's argument in *Phaedrus* can easily be countered, and the same applies to its echoes in modern times. What both of these versions of the 'extension argument' overlook, is the fact that technological artefacts, though they are not part of our organic body, are an integral part of our distributed cognitive structure (cf. Magnani 2007: 5–6). They remain part of ourselves, as the artefacts are part of the conjoint network in which we operate and act with them. Moreover, what the extension argument misses is the fact that technological artefacts, in adopting and reconfiguring certain tasks from human beings, facilitate the development and blooming of all kinds of new 'typically human' capacities. By delegating the content of our memories to paper (in writing), our cognitive structure is less burdened with the task of remembering, and thus new roads are opened for the development of novel forms of rationality, structured by the medium-specific characteristics of writing. This same mechanism applies to delegating the process of rational thinking to computer technologies. Such delegation does not lead to a diminishment of human autonomy, but to an increase of human agency, and as such, to an expansion and strengthening of human autonomy. In a sense the more agency an artefact has, the more it potentially *enhances* human autonomy by inviting us to reach new goals and use new means.

The critique of Plato and his modern heirs starts from a dichotomous distinction between human agents and technological artefacts. This distinction is problematic, because human beings and artefacts have always and will always form *networks*, in which each mutually depends on the other (Latour 1993, 1999, 2005, Magnani 2007). Neither can exist without the other – a human being is not a human being without artefacts, nor is an artefact an artefact without human beings.

When we *do* distinguish between human beings on the one hand and artefacts on the other (either analytically or in practice), claiming that human beings are active whereas artefacts are passive is an obvious oversimplification. As we have argued in the preceding section, human beings have never been very autonomous. A considerable part of our actions is remote controlled by both internal and external factors that are outside our sphere of control. Our human agency is not a completely autonomous (*self-governing*) power, but rather a reflexive relation to that which motivates our actions – a relation, moreover, in which we can choose to affirm and absorb these motivating forces as *our* motives, drives, passions, ideas. Only in the interplay of our internal and external motivators on the one hand and our own reflexive appropriation on the other do we, as acting beings, as agents, emerge. Human agency, in this sense, has always been distributed agency.

Bruno Latour has raised a similar argument regarding the moral implications of technological mediation. Latour rejects the assumption that human ethics formulates moral goals, whereas technology merely supplies the means to realize these goals. Technology always provides us with a *detour* towards the goals we want to reach:

> If we fail to recognize how much the use of a technique, however simple, has displaced, translated, modified, or inflected the initial intention, it is simply because we have *changed the end in changing the means*, and because, through a slipping of the will, we have begun to wish something quite else from what we at first desired. [. . .] Without technological detours, the properly human cannot exist. [. . .] Technologies bombard human beings with a ceaseless offer of previously unheard-of positions – engagements, suggestions, allowances, interdictions, habits, positions, alienations, prescriptions, calculations, memories.
>
> (Latour 2002: 252)

Beyond human agency

In the traditional conception of technologies, one could argue, we conceive of our own relationship towards these technologies as follows: we, as human beings, formulate one or more *goals* or outcomes we want to achieve, and we then proceed to create technologies to reach those goals. We are in charge of the outcomes of technologically mediated praxes, and we provide the technology with the *processes* to go about reaching the goal we have set. Technologies are thus viewed as simple instruments, with which we have a clear goal-means relationship.

In fact, our relationship with technologies is much more complex and diversified than that (cf. De Mul 2009: 245–61). While we do in fact create *some* technologies for which we define both the goals and the processes of reaching those goals, there are also examples of technologies for which we define *only* the outcomes. The process of reaching those outcomes is left to the artefact itself. For instance, in modern cars, when we press the brake to make it stop, the brake system, consisting of independently operating sub-systems, cleverly 'decides'[4] which systems it needs to engage in those specific circumstances to make the car stop. Moreover, in some cases not only the *process* of accomplishing certain goals is left to the technology, but the definition of the *outcome* itself as well. Both the goals and the process are thus delegated to the technology. This is the case, for instance, in the 'power grid', a network of power supply systems that manage the power distribution in Southern California. The grid decides how to distribute power optimally (process), but also defines what the best outcomes of distribution are (goal). It is clear that all three of these forms entail different relationships between human beings and technological artefacts, and have consequences for human autonomy.

At the beginning of this chapter we discussed our recent voting experiences surrounding the elections for the Water Boards in the Netherlands. We described the use of the Electoral Compass, designed and deployed to relieve us of the burden of having to muddle our way through large amounts of electoral information, ranging from flyers and websites to television debates and party programmes. We concluded that what happens when we use a technology such as the Electoral Compass is a delegation of the *process* to the technology, whereas the *goal* – that is, the final decision of who to vote for and the actual casting of the vote – remains firmly in the hands of the human agent. It is the individual, autonomous agent, who uses the information provided by the technology, but weighs and decides for himself. In the case of the Electoral Compass, therefore, we would be hard pressed to argue that using this technology undermines our autonomy.

As we have seen, one could argue that matters are a little different in the case of one of the other voting aides we discussed. In the case of VoteMatch, the technology does not provide us with an overview in numbers of our alignment to each of the parties we can vote for, but with an explicit *advice* instead. The technology clearly points us in the direction of a specific vote, and this raises the question of influencing. How many of us would be recalcitrant and daring enough to ignore the advice and vote for an entirely different party? How many of us would be aware of the fact that they do not *have* to accept the advice? How many of us would think critically about either the content of the advice, or the ways in which it has been constructed? We have seen above how many faceted the scripting of this kind of technologies really is. One could argue, therefore, that in this second case the decision to vote for this party, rather than that one, is indeed delegated to the technological artefact in a way that diminishes (although it doesn't eradicate) our human autonomy.

Now, think of the following scenario. What if, in the near future, a voting aide would be able not only to provide us with an overview of our political alignment to various parties, or with a clear and concise, boxed-in, ready-made voting advice, but would also be able to take things one step further? What if the voting aide could also cast the vote for us? Imagine that we, as autonomous agents, would be too busy, too lazy, too cynical or otherwise engaged in any other form or shape, to put forth the effort of actually casting the vote on the designated election day. What would that entail for our autonomy? If the voting aide would consult us in a final decision on who to vote for before it actually casts the vote, this seems fair enough. After all, it doesn't actually matter (or does it?) who presses the button on the voting machine, or who handles the red pencil to fill out the form. What matters is that *my* decision as a voter (which, notably, I may have come by through my own rational or not-so-rational electoral choices, or with the more or less neutral help of a voting aide) is the final decision in this fundamental democratic practice. My human autonomy (or what is left of that after accepting VoteMatch's advice as *my* voting decision) is safeguarded as long as I have the final say.

But what if we take it yet another step further? Imagine a world in which the voting aide does all of the things discussed before, but on top of that it can also vote for us *in an independent way*. This means, for instance, that despite the fact that we have always voted for party X (and may even have told the aide explicitly to do so, this time, for us as well), it may decide that it has good grounds to ignore our voting history and our current voting preference, and to vote for an entirely different party *in our name*. Let's assume that it doesn't choose to do so out of spite or confusion or any 'irrational' or non-benevolent motivation, but that it merely uses profiling to calculate that, even though we always thought that X was the party for us, in reality our ideas and our behaviours match Y much more closely, and therefore Y is the party we *should* vote for. In this scenario Y is not only the party we *should* vote for, but the one we effectively *do* vote for, because the voting aide will cast the vote on our behalf to its best judgement and without our explicit consultation. It is obvious that, in this scenario, our human agency is indeed seriously affected by the process of delegation to a technological artefact, and that our remote control is undeniably extremely remote in this case.

Now, it is easy to cast aside a scenario such as this with the argument that it is futuristic (ergo unrealistic) 'what-if babble'. However, look at the following scenario, which we have clipped from one of the key documents presenting the European Commission's perspective of the near technological future, called Ambient Intelligence:

> It is four o'clock in the afternoon. Dimitrios, a 32 year-old employee of a major food-multinational, is taking a coffee at his office's cafeteria, together with his boss and some colleagues. He doesn't want to be excessively bothered during this pause. Nevertheless, all the time he is receiving and dealing with incoming calls and mails. [. . .] Dimitrios is wearing, embedded in his clothes [. . .], a voice activated 'gateway' or digital avatar of himself, familiarly known as 'D-Me' or 'Digital Me'. A D-Me is both a learning device, learning about Dimitrios from his interactions with his environment, and an acting device offering communication, processing and decision-making functionality. Dimitrios has partly 'programmed' it himself, at a very initial stage. [. . .] He feels quite confident with his D-Me and relies upon its 'intelligent' reactions.
>
> At 4:10 p.m., following many other calls of secondary importance – answered formally but smoothly in corresponding languages by Dimitrios' D-Me with a nice reproduction of Dimitrios' voice and typical accent, a call from his wife is further analysed by his D-Me. In a first attempt, Dimitrios' 'avatar-like' voice runs a brief conversation with his wife, with the intention of negotiating a delay while explaining his current environment. [. . .] [However, when she calls back once more] his wife's call is [. . .] interpreted by his D-Me as sufficiently pressing to mobilise Dimitrios. It 'rings' him using a pre-arranged call tone. Dimitrios takes up the call with one of the available Displayphones of the cafeteria. Since the growing penetration of D-Me, few people still bother to run around with mobile terminals: these functions are sufficiently available in most public and private spaces [. . .] The 'emergency' is about their child's homework. While doing his homework their 9 year-old son is meant to offer some insights on everyday life in Egypt. In a brief 3-way telephone conference, Dimitrios offers to pass over the query to the D-Me to search for an available direct contact with a child in Egypt. Ten minutes later, his son is videoconferencing at home with a girl of his own age, and recording this real-time translated conversation as part of his homework. All

communicating facilities have been managed by Dimitrios' D-Me, even while
it is still registering new data and managing other queries.

(Ducatel et al. 2001: 5)

This scenario is not about a voting aide, nor about our actions as autonomous voting
agents, but it does portray a number of relevant parallels with the last stage of
the voting aide we have discussed above. The man in the scenario has a personal
technological aide that answers his incoming communications whenever he is
otherwise occupied. Although this sounds quite appealing, and not even so uncom-
mon at first – most of us use answer phones and automatic e-mail replies for
precisely the same goal – there are two rather eerie elements to his aide's capacities
and behaviours. First, the aide has been given the responsibility to decide whether
incoming messages are important or not. Note that an answer phone or an automatic
e-mail reply are entirely indiscriminate in this respect. The aide thus makes
decisions based on *its* estimate of the importance of the content of the message and
the nature of the relationship one has to the caller. This means that it *values* our
communications for us *and acts* on the basis of these values. Second, what is eerie
about the aide in this example is that it *mimics its owner*. It responds to incoming
communications using an imitation of its owners' voice, including inflections and
word choice. This means that we do not only delegate the process of valuing to the
artefact, but also the *form and content of our* [sic] *response*. And these are precisely
the same two issues that are at stake in the scenario we've sketched for the voting
aide of the future.

Delegating agency to artefacts is something human beings have been prone to
do since the beginning of time. No harm is done in most of our delegations – quite
the reverse. They enhance our abilities to act in the world and create new possibilities for action that would be impossible without such delegation. With the
advent of autonomic computing and Ambient intelligence the delegation of agency
reaches hitherto unimaginable levels, and our degree of 'competence', effectiveness
and autonomy will thereby stretch to new limits. This is why these technological
developments deserve our support and attention. But at the same time, we must
always be vigilant of the turning point, at which the autonomy and agency of human
agents are externalized to such a degree that they in fact undermined considerably.
This means that the challenge for both designers and social scientists and philosophers is to find this turning point, to approach it as closely as possible, yet to never
cross it.

We argue that the reflexive loop that we have discussed in this chapter is crucial
in this respect. The danger is not so much in delegating cognitive tasks, but in
distancing ourselves from – or in not knowing about – the nature and precise
mechanisms of that delegation. As we noted in our discussion of the voting aides
artefacts contain scripts on two different levels: they contain various (technological and political) ideas and norms from the designers who built them, and they
influence users' thinking and actions. Awareness of, and insight into the 'scriptal
character' of the artefact, and having the ability to influence that character, is crucial

for users in light of the delegation of their autonomy. If we lack awareness and insight with respect to the way a voting aide works, the 'prejudices' that it (unavoidably) contains, and the grounds on which 'our' choice is made, then our autonomy is threatened, even if this choice is in line with our political preferences and interests. If we do have this awareness and insight, and a reflexive loop enables us to toy with the aforementioned parameters and to confirm or reject certain values, then the knowledge and decision rules that are built into the voting aide will strengthen our autonomy instead. In that case distributed agency entails an enhancement of our power to act.

Of course, human awareness and knowledge are limited. As computer systems become more and more complex, it will be ever more difficult to open and understand the black box. It is likely, therefore, that the reflexive loop will gradually move from the organic to the artificial components of the network to an ever larger degree. Conceivably, such 'intentional networks' will be superior to networks in which the human 'component' is the final link. From an anthropocentric perspective that is quite something. Yet it would be unwise to follow Medea and kill our mind children, the technological artefacts, based on hurt pride. Instead, maybe we can find comfort in these words, uttered by Nietzsche's Zarathustra:

> Man is a rope, tied between beast and Overman – a rope over an abyss [. . .]
> What is great in man is that he is a bridge and not an end: what can be loved
> in man is that he is an *overture* and a *going under*. . .
>
> (Nietzsche 1980, Vol. 4: 16)

Notes

1 For instance, in 2006, a year of national elections in the Netherlands, 64 per cent of the voters expressed that they had used none of the sources of information discussed here to inform themselves before casting their vote (CBS 2006).

2 'Agency' here is to be understood as the 'capacity to act', whereby we leave open the question of the precise necessary and/or sufficient conditions for such a capacity to arise. The 'standard conception' of agency summarizes the notion of agency in the following proposition: 'X is an agent if and only if X can instantiate intentional mental states capable of directly causing a performance.' (Himma 2008: 3). However, this entails a discussion of what intentionality is, and which beings qualify as 'really' intentional – as Daniel Dennett has remarked 15 years ago already: '. . .for the moment, let us accept [the] claim that no artifact of ours has real and original intentionality. But what of other creatures? Do dogs and cats and dolphins and chimps have real intentionality? Let us grant that they do; they have minds like ours, only simpler, and their beliefs and desires are as underivedly about things as ours. [. . .] What, though, about spiders or clams or amoebas? Do they have real intentionality? They process information. Is that enough? Apparently not, since computers – or at least robots – process information, and their intentionality (*ex hypothesi*) is only derived.' (Dennett 1994: 100). We follow Dennett in his solution to the intentionality question: what matters is not so much whether an organism has intentionality or not, but whether it displays something that convinces us as being intentionally aimed – Dennett calls this '*as-if* intentionality' (cf. Adam 2008, Dennett 1994). Moreover, in our conception of agency, we side with Floridi and

Sanders, who formulate three criteria for agency: (1) interactivity, (2) autonomy, and (3) adaptability (Floridi and Sanders 2004: 349, 357–58).

3   We deliberately put the phrase 'made a decision' between quotation marks here to indicate that we should take this phrase as a *façon de parler*. Although many contemporary neuroscientists ascribe psychological attributes (such as making decisions) to the brain, this should be regarded as a category mistake if this is taken literary and not as a metaphor. After all, brains do not make decisions, only human beings do. Neuroscience can investigate the neural preconditions for the possibility of the exercise of distinctively human powers such as thought, reasoning and decision-making and discover correlations between neural phenomena and the possession (or lack) of these powers, but it cannot simply replace the wide range of psychological explanations with neurological explanations. When neuroscientists ascribe psychological attributes to brains instead of to the psychophysical unity that constitutes the human being, they remain victims of (an inverted version of) Cartesian dualism (cf. Bennett 2007: 6–7, 142ff., Bennett and Hacker 2003: introduction). The fact that neuroscientific investigations show that (in specific cases) neural processes that accompany bodily action precede conscious decision, does not prove that the brain makes the decision instead, but rather that in these cases the psychophysical unity decides unconsciously.

4   Here, too, we have placed the word 'decides' between quotation marks, since the brake system does not decide in the ordinary sense of the word, but rather acts mechanically according to its programme. The point is, however, that the brake system functions independently of the driver. The more complicated an automated device, the more we will tend to ascribe intentionality and even rational decision-making to it.

References

Adam, A. (2008) 'Ethics for things', *Ethics and Information Technology,* 10: 149–54.

Akrich, M. (1992) 'The de-scription of technical objects', in Bijker, W. E. and Law, J. (eds) *Shaping technology/building society: Studies in sociotechnical change.* Cambridge, MA: MIT Press.

—— (1995) 'User representations: Practices, methods and sociology', in Rip, A., Misa, T. J. and Schot, J. (eds) *Managing technology in society: The approach of constructive technology assessment.* London, New York: Pinter Publishers.

Alford, C. F. (1992) 'Responsibility without freedom. Must antihumanism be inhumane? Some implications of Greek tragedy for the post-modern subject', *Theory and Society,* 21: 157–81.

Bennett, M. R. (2007) *Neuroscience and philosophy: Brain, mind, and language.* New York: Columbia University Press.

Bennett, M. R. & Hacker, P. M. S. (2003) *Philosophical foundations of neuroscience.* Malden, MA: Blackwell Pub.

Berg, A.-J. (1999) 'A gendered socio-technical construction: The smart house', in MacKenzie, D. A. & Wajcman, J. (eds) *The social shaping of technology.* 2nd ed. Buckingham (UK), Philadelphia (PA): Open University Press.

Burns, K. & Bechara, A. (2007) 'Decision making and free will: A neuroscience perspective', (25) *Behavioral Sciences & the Law,* 2: 263–80.

CBS (2006) 'Politieke en sociale participatie'. CBS.

Dawkins, R. (2006) *The selfish gene.* Oxford: Oxford University Press.

De Mul, J. (2003) 'Digitally mediated (dis)embodiment: Plessner's concept of excentric positionality explained for cyborgs', (6) *Information, Communication & Society,* 2:

247–66.

—— (2004) *The tragedy of finitude: Dilthey's hermeneutics of life*. New Haven: Yale University Press.

—— (2009) [2006] *De domesticatie van het noodlot: De wedergeboorte van de tragedie uit de geest van de technologie*. Kampen (NL), Kapellen (Belgium): Klement/ Pelckmans.

Dennett, D. C. (1994) 'The myth of original intentionality', in Dietrich, E. (ed.) *Thinking computers and virtual persons: Essays on the intentionality of machine*. San Diego: Academic Press.

Dilthey, W. (1914–2005) *Gesammelte Schriften (23 vols.)*. Stuttgart/Göttingen: B.G.Teubner, Vandenhoeck & Ruprecht.

Ducatel, K., Bogdanowicz, M., Scapolo, F., Leijten, J. & Burgelman, J.-C. (2001) 'ISTAG: Scenarios for Ambient Intelligence in 2010'. Seville (Spain): IPTS (JRC).

Ellul, J. (1988) *Le bluff technologique*. Paris: Hachette.

Euripides, (2006) *Medea*. Oxford; New York: Oxford University Press.

Floridi, L. & Sanders, J. W. (2004) 'On the morality of artificial agents', *Minds and Machines*, 14: 349–79.

Gergen, K. J. (2002) 'The challenge of absent presence', in Katz, J. E. & Aakhus, M. A. (eds) *Perpetual contact: Mobile communication, private talk, public performance*. Cambridge (UK), New York (NY): Cambridge University Press.

Gjøen, H. & Hård, M. (2002) 'Cultural politics in actions: Developing user scripts in relation to the electric vehicle', (27) *Science, Technology & Human Values*, 2: 262–81.

Gontier, T. (2005) *Descartes et la causa sui: Autoproduction divine, autodétermination humaine*, Paris: J. Vrin.

Heidegger, M. (1962) *Die Technik und die Kehre*, Pfullingen: Neske.

Himma, K. E. (2008) 'Artificial agency, consciousness, and the criteria for moral agency: What properties must an artificial agent have to be a moral agent?', (11) *Ethics and Information Technology*, 1: 19–29.

Latour, B. (1992) 'Where are the missing masses? The sociology of a few mundane artifacts', in Bijker, W. E. & Law, J. (eds) *Shaping technology/building society: Studies in sociotechnical change*. Cambridge, MA: MIT Press.

—— (1993) *We have never been modern*. Cambridge, MA: Harvard University Press.

—— (1999) *Pandora's hope: Essays on the reality of science studies*. Cambridge, MA: Harvard University Press.

—— (2002) 'Morality and technology: The end of the means', (19) *Theory, Culture & Society*, 5–6: 247–60.

—— (2005) *Reassembling the social: An introduction to actor-network-theory*. Oxford (UK), New York (NY): Oxford University Press.

Libet, B. (1985) 'Unconscious cerebral intitiative and the role of conscious will in voluntary action', (8) *Behavioral and Brain Sciences*, 4: 529–66.

Magnani, L. (2007) 'Distributed morality and technological artifacts', paper presented at Human Being in Contemporary Philosophy, Volgograd (Russia), 28–31 May.

Nietzsche, F. (1980) *Sämtliche Werke (15 vols.)*. Berlin: De Gruyter.

Ong, W. J. (1982) *Orality and literacy: The technologizing of the word*. London, New York: Methuen.

Plato (1914) *Plato, with an English Translation* (trans. North, H., Fowler, L. & Maitland, W. R.). London, New York: W. Heinemann, The Macmillan Co.

Plessner, H. (1981) *Die Stufen des Organischen und der Mensch: Einleitung in die philosophische Anthropologie*. Frankfurt am Main: Suhrkamp.

Rimbaud, A. J. (1871) *Lettre du Voyant*, in a personal communication to Demeny, P., 15 May 1871.

Rosenberg, A. (2008) *Philosophy of social science*. Boulder, CO: Westview Press.

Schank, R. C. & Abelson, R. P. (1977) 'Scripts', *Scripts, plans, goals, and understanding: An inquiry into human knowledge structures*. Hillsdale (NJ), New York (NY): L. Erlbaum Associates.

Van den Berg, B. (2009) *The situated self: Identity in a world of Ambient Intelligence*. Rotterdam: Erasmus University.

Van Oost, E. (2003) 'Materialized gender: How shavers configure the users' femininity and masculinity', in Oudshoorn, N. and Pinch, T. J. (eds) *How users matter: The co-construction of users and technologies*. Cambridge, MA, MIT Press.

Chapter 4

Autonomy, delegation, and responsibility

Agents in autonomic computing environments

Roger Brownsword[1]

Introduction

Suppose that our everyday environments were to be technologically enabled so that they became smarter, more intelligent, more anticipatory, and more responsive to our needs? Imagine 'a world of convergence, where heterogeneous devices are able to communicate seamlessly across today's disparate networks, a world of machine learning and intelligent software, where computers monitor our activities, routines and behaviours to predict what we will do or want next' (Wright et al. 2008: 1).[2] This is a world that gives rise to a host of concerns, to concerns about privacy, security, trust, and reliability, and the like. It is also a world in which—as Jos de Mul and Bibi van den Berg (2011) have vividly highlighted—there are concerns about whether there is a future for agents, particularly whether there is any room for agent autonomy and any role for agent responsibility.

More precisely, de Mul and van den Berg's paper invites a response in relation to the following three key questions (concerning, respectively, autonomy, delegation, and responsibility):

(1) What are the implications for agent *autonomy* in autonomic computing environments (ACEs)?
(2) What are the implications of agents *delegating* their decision-making to smart computing aides?
(3) What is left of agent *responsibility* in ACEs?

At one level, my response to these questions will be in the nature of a mapping and classificatory exercise. It will highlight three conceptions of autonomy, various degrees of delegation, and two types of responsibility, relative to which we can categorise particular design features as 'enhancing', 'eroding', or 'eliminating' in relation to specified values (for example, in relation to values such as autonomy, privacy, and dignity). At another level, however, my response will move beyond such classificatory questions to consider how ACEs impact on the aspirations that agents have for the creation of moral communities (particularly for communities of rights). Without being unduly alarmist, the paper will address the concerns that such agents might have not just about the disruptive effects of new technology (see, e.g., Klang 2006) but about the survival of their communities.

The paper is in four principal parts. In the first three of these parts, I conduct the mapping and classifying exercise, focusing, respectively, on autonomy, delegation and responsibility. In the fourth part, I offer some short thoughts about the prospects for agents in a particular kind of aspirant moral community, a community of rights.

What are the implications for agent autonomy in an environment of autonomic computing?

Before we can assess the implications of ACEs in relation to agents and their autonomy, we need to be clear about how we understand agency and autonomy. Accordingly, in the first section of this part of the paper, having characterised agency in terms of the capacity for free and purposive action, I identify three guiding (but competing) conceptions of autonomy; and, then, having incorporated what I take to be de Mul and van den Berg's gloss on free choice, I sketch some ways in which, relative to each conception, autonomic computing technologies might compromise autonomy.

Conceptions of agency and autonomy

Let us suppose that we treat X as an agent where we have reasonable grounds to believe that X has the capacity to act freely and purposively. This does not mean that X always acts freely; and nor does it mean that X's actions are always purposive. Nevertheless, the working assumption in both legal and moral discourse is that the addressees of practical prescriptions have the capacity to respond freely and purposively. Put more directly, we can simply say that X is an agent where X has the capacity to act freely and purposively; and that X acts as an agent where X acts freely and purposively.

If we work with this understanding of agency, the next question is how we conceive of autonomy. Under what conditions would we say that agent X acts autonomously? From the many possible responses to this question,[3] let me single out the following three:

(a) Agent X acts autonomously when X acts for a purpose, P, that is freely chosen by X (in the sense that X chooses P without being subject to external pressure and influence).
(b) Agent X acts autonomously when (i) X acts for a purpose, P, that is freely chosen by X (in the sense that X chooses P without being subject to external pressure and influence) and (ii) purpose P is consistent with X's stable preferences, longer term interests, or critical interests.[4]
(c) Agent X acts autonomously when (i) X acts for a purpose, P, that is freely chosen by X (in the sense that X chooses P without being subject to external

pressure and influence) and (ii) purpose P is consistent with moral reason and requirements.

In each case, the devil can lie in the detail of these responses. For example, the provision (common to each conception) that X's choice of P should be 'free' is open to many different interpretations. If we judge that, in organised societies, it is unrealistic to expect our actions to be wholly insulated against external pressure or influence—if we judge that, even though the last man or woman standing might enjoy such freedom, it makes no sense to specify autonomy in a way that imposes such a demanding requirement—then we are likely to settle for a weaker version of freedom. For instance, we might weaken the requirement of free choice so that X's choice must be 'substantially' or 'reasonably' free; or we might say that the choice must at least be made without 'undue' external pressure or influence, or that there should be no 'abnormal' or 'exceptional' pressure, or something of that kind. Once we move the spotlight away from the requirement of voluntariness to that of purposivity, we again find scope for various interpretations—for example, in the second conception, the idea of an agent's 'critical' interests is open to a number of readings; and, in the third response, the requirement that P should be consistent with moral requirements is open to as many interpretations as there are criteria of moral action.

One of the principal claims made by de Mul and van den Berg is precisely that it is unrealistic to conceive of agents as entirely free and independent authors of their own choices: social life simply is not like that. It follows that any account of autonomy must recognise this social reality. The only question then is how far the requirement of free choice is to be weakened. De Mul and van den Berg propose that it is sufficient that agent X acts in a way that is self-consciously at a distance from the pressures, influences, and restrictions that make up the context for action. The range of choices available to agents, like the number of available TV channels, varies from one situation to another; for agent X to act autonomously, the final choice must lie with X—X must not, as it were, surrender the remote control. For present purposes, and for the sake of argument, we can adopt this proposed specification of free choice which, when copied into the responses above, yields the following:

(aa) Agent X acts autonomously when X acts for a purpose, P, that is freely chosen by X (in the sense that X chooses P in a way that is self-consciously at a distance from the pressures, influences, and restrictions that make up the context for action).

(bb) Agent X acts autonomously when (i) X acts for a purpose, P, that is freely chosen by X (in the sense that X chooses P in a way that is self-consciously at a distance from the pressures, influences, and restrictions that make up the context for action) and (ii) purpose P is consistent with X's stable preferences, longer-term interests, or critical interests.

(cc) Agent X acts autonomously when (i) X acts for a purpose, P, that is freely chosen by X (in the sense that X chooses P in a way that is self-consciously at a distance from the pressures, influences, and restrictions that make up the context for action) and (ii) purpose P is consistent with moral reason and requirements.

Although, by adopting this proposal, we stabilise the voluntariness requirement for autonomous action, we still have significant differences with regard to the element of purposivity. Whether or not this matters rather depends upon our particular cognitive interest.

If our cognitive interest is in mapping the way in which autonomic computing environments might impinge upon the possibility of X choosing P in a way that is self-consciously at a distance from the pressures, influences, and restrictions that make up the context for action, it matters little which of the three conceptions we operate with. For, the question is the same irrespective of the conception that is employed. This also holds good if our cognitive interest is not simply in mapping per se but in holding on to practices that keep the technology at arm's length from agents. However, if our cognitive interest is in retaining and promoting practices that specifically advance the 'real' interests of agents (whether those interests are taken in a prudential or a moral sense), then our inquiry will be driven by a more particular conception of autonomy, that is, by a conception that is closer to the second (bb) or to the third account (cc) than to the first (aa).

Given these clarificatory remarks we can move on to the issue of the compatibility between various conceptions of autonomy and ACEs, or decision aides, or the like.

Autonomy and autonomic computing environments

If we focus on conceptions (aa), (bb), and (cc), what kind of ACEs, or decision aides, would be incompatible with agent X acting autonomously?

(aa) For agent X to act autonomously X must act for a purpose, P, that is freely chosen by X (in the sense that X chooses P in a way that is self-consciously at a distance from the pressures, influences, and restrictions that make up the context for action).

Given this conception of autonomy, agent X would *not* act autonomously if:

- The environment or aides do not allow for X to act at a distance from the selected or recommended or advised option (or)
- The environment or aides do allow for X to act at a distance from the selected or recommended or advised option, but X does not actually act a distance.

On this view, it is important that ACEs are designed in such a way that (i) they leave open the possibility of autonomous action and (ii) they guard against undue influence—possibly by requiring some active step by X to adopt the recommended or advised option.[5] Just as technology designs should be privacy enhancing, they should also be autonomy enhancing (AED). For, just as (say) surveillance and tracking technologies might inhibit deviation from standard social norms or mainstream patterns of conduct (Griffin 2008), ACEs might act as an invisible hand guiding agents towards conformist patterns of behaviour.

Consider the case of on-line behavioural advertising (EC Health and Consumers Directorate-General 2009). Where consumers are persistently presented with advertising that (accurately) reflects their previous pattern of consumption, is there a danger that this not only militates against experimentation but also makes it difficult to keep profiling technology at a distance? Suppose that the point were to be tested legally: might behavioural advertising be considered as a form of undue influence under Directive 2005/29/EC, the Unfair Commercial Practices Directive (UCPD)? According to Article 2(j) of the UCPD, '"undue influence" means exploiting a position of power in relation to the consumer so as to apply pressure, even without using or threatening to use physical force, in a way which significantly limits the consumer's ability to make an informed decision.' Section 2 of the UCPD sets out the substantive regulatory provisions with regard to aggressive commercial practices. Article 8 treats a practice as aggressive if

> In its factual context, taking account of all its features and circumstances, by harassment, coercion. . .or undue influence, it significantly impairs or is likely to impair the average consumer's freedom of choice or conduct with regard to the product and thereby causes him or is likely to cause him to take a transactional decision that he would not have taken otherwise.

Article 9 then sets out the indicators of harassment, coercion or undue influence, starting with the 'timing, location, nature, or persistence' of the practice; and Annex I to the Directive specifies eight aggressive practices that are to be treated as unfair in all circumstances. Insofar as these eight instances throw light on the relevance of 'timing, location, nature, or persistence', we might note that one case is that of 'Making persistent and unwanted solicitations by telephone, fax, e-mail, or other remote media. . . .'

Speaking as a common lawyer, the UCPD definition of undue influence is doubly surprising: first, in its emphasis on the exploitative behaviour of the commercial party (in English law, the better view is that undue influence does not imply any wrongdoing; the question is whether the consumer's power of autonomous decision-making has been compromised)[6]; and, secondly, in its focus on whether the consumer's ability to make an *informed* [sic, rather than a *free*] decision has been compromised. However, there is no real problem with this: the definition is for the purposes of the UCPD only; and, by the time that we get into the substantive provisions of the Directive, undue influence seems to have been more accurately allocated (along with harassment and coercion) to aggressive practices where the issue is about the consumer's ability to make a free decision.

The argument that behavioural advertising might amount to a form of undue influence supposes that the effect of the advertising is to limit the consumer's horizons—it is almost as though the consumer gets trapped by his or her previous retailing record. Imagine that you are a regular customer at an Italian restaurant where you always eat a particular pasta. As soon as you sit at your favourite table, the waiter asks, 'The usual pasta, sir?' Not wishing to disrupt this practice, you go along with it, never exploring the menu beyond the range of pasta choices. In an off-line context, would this amount to undue influence? Now imagine that this is what happens in an ACE where the technology pretty much has your pasta on the table as soon as you enter the restaurant. Would this amount to undue influence?

There are at least two important qualifiers in the UCPD. First, the limitation or impairment of the consumer's decision-making power has to be significant (see both Article 2(j) and Article 8). Secondly, according to Article 8, the benchmark is the 'average consumer', a standard that is open to considerable interpretation but which clearly is not pitched at the most vulnerable of consumers. Moreover, there seems to be a need not only for exploitative conduct by the commercial party (a characterisation that might not fit smoothly with ACEs)[7] but also for the making of a decision by the consumer that otherwise would not have been made. While we cannot be entirely confident about the way that the UCPD might be applied or interpreted in relation to behavioural advertising, it seems unlikely—particularly bearing in mind that the average consumer sets the standard and that, presumably this consumer is exposed to a range of advertising in both on-line and off-line milieu—that such practices would normally constitute undue influence.

Nevertheless, even if this does not amount to 'undue influence' within the meaning of the UCPD, it might be sufficient to raise a question about the autonomy of the consumer's choice. The test proposed by de Mul and van den Berg, let us recall, is that X chooses P in a way that is self-consciously at a distance from the pressures, influences, and restrictions that make up the context for action. Where X chooses in an ACE, and where agents are subject to saturation behavioural advertising, the selections made by consumers might be too close to the technological prompts for comfort. As with the legal position under the UCPD, the philosophical and psychological position concerning the survival of autonomy in settings that are geared for behavioural advertising surely is moot.

Before we continue, it should be emphasised that none of this addresses the possibility that the agent self-consciously delegates the final decision to the technology. In such a case, the delegating agent autonomously waives the option of keeping the technology at a distance; and this might be seen as compatible with autonomous action. Accordingly, the comments above must be taken as applying only to a context in which there is no explicit and self-conscious delegation to the technology of final decision.

(bb) Agent X acts autonomously when (i) X acts for a purpose, P, that is freely chosen by X (in the sense that X chooses P in a way that is self-consciously at a distance from the pressures, influences, and restrictions that make up the context for action) and (ii) purpose P is consistent with X's stable preferences, longer-term interests, or critical interests.

On this view of autonomy, as in (aa), agent X would *not* act autonomously if:

- The environment or aides do not allow for X to act at a distance from the selected or recommended or advised option (or)
- The environment or aides do allow for X to act at a distance from the selected or recommended or advised option, but X does not actually act a distance.

However, in conception (bb), there is a degree of discrimination about the purposes that X chooses. To act autonomously, X must choose a purpose that is, broadly speaking, in line with X's own interests. Accordingly, under conception (bb), the important additional requirement is that agent X would *not* act autonomously if:

- The freely chosen purpose, P, is not consistent with X's stable preferences, longer-term interests, or critical interests.

Given such a conception of autonomy, the novel design challenge is to develop ACEs that correct for inconsistency on the part of the agent or steer agents towards purposes that are conformist relative to the agent's own preferences or interests.

Consider the employment of a voting-aide in the kind of scenarios sketched by de Mul and van den Berg. To start with a simple case, suppose that the technology is fairly primitive, and that it simply reminds agent X that, having in the past consistently voted for one party (the Greens, let us say), it would now be consistent for X to cast a vote once again for the Greens. As the technology becomes more sophisticated it is able to advise X that the Greens' manifesto still fits best with X's core beliefs; and, in one of the scenarios, the voting aide advises that, given the full set of X's beliefs and behaviour, it would be more consistent for X to vote for another party (the Reds, let us say).

As we have remarked already, the devil is in the detail of these competing conceptions of autonomy and their applications in ACEs. As the voting aide technology becomes more complex in its capacity to confirm or correct X's preferences or interests, we see the problematic nature of judging that X acts autonomously relative to conception (bb). Where the technology, at an early stage of its development, simply confirms X's apparent preferences, all seems to be well—or, at any rate, all seems to be well unless we pause to ask whether X's apparent preferences are X's 'real' preferences. However, as the technology becomes more sophisticated, as it develops the capacity to take a broader and deeper look at X's beliefs and behaviour, we begin to see that X's apparent preferences might not reflect X's real interests. At this stage, we cannot salvage things by staying with the simpler technology—because this might be advising X in ways that do not represent his real interests; in which case, the technology is guiding X in ways that lead him away from autonomous acts. On the other hand, the more complex algorithms that drive the more sophisticated voting aides will themselves be contestable—because the harder we look into X's real interests, the harder it will be to stabilise our understanding and application of this version of autonomy.

There is a further issue. So far, we have presupposed that the voting aide operates only in an advisory way: the voting aide advises X that he should vote for the

Greens (which is the advice that X expects to receive); or it advises that X should vote for the Reds (which is not the advice that X expects to receive). Whatever the advice, though, it is no more than advice. If the technology goes beyond advice to actually cast the vote on X's behalf, is this problematic?

Intuitively, we might think that, while the technology is not problematic where the vote that is cast (for the Greens) is the vote that X expects to be cast, there is a problem where the vote that is cast (for the Reds) is not the vote that X expects to be cast. Yet, if, in both cases, the vote that is cast is actually in line with X's real interests, this seems to be an autonomous act by X (assuming, I should hasten to add, that there is no problem here about X still acting 'freely').[8] Nevertheless, as a precautionary measure, we might think it important that, in the latter case, the unexpected vote for the Reds should not be automatically cast. In this case, where X is not aware of this inconsistency in his actions (relative to the full range of his preferences), we might want the technology to alert him to the discrepancy rather than simply taking what seems to be the appropriate corrective action. After all, X's preferences have a dimension of weight, as well as (so to speak) a 'provisionality' (that is, they are always subject to revision). Accordingly, we might conclude that even the most sophisticated voting aide (relative to this particular conception of autonomy) should be designed in such a way that the matter is referred back to X for a final decision. Furthermore, other things being equal, such a design would improve the prospects of X acting at a distance from the technology.

(cc) Agent X acts autonomously when (i) X acts for a purpose, P, that is freely chosen by X (in the sense that X chooses P in a way that is self-consciously at a distance from the pressures, influences, and restrictions that make up the context for action) and (ii) purpose P is consistent with moral reason and requirements.

On this view of autonomy, as in both conceptions (aa) and (bb), agent X would *not* act autonomously if:

- The environment or aides do not allow for X to act at a distance from the selected or recommended or advised option (or)
- The environment or aides do allow for X to act at a distance from the selected or recommended or advised option, but X does not actually act at a distance.

However, in conception (cc), as in (bb), there is a discriminating approach to the purposes that X chooses. To act autonomously, X must choose a purpose that passes moral muster. Accordingly, under conception (cc), the important additional requirement is that agent X would *not* act autonomously if:

- The freely chosen purpose, P, is not consistent with moral reason and requirements.

Given such a conception of autonomy, the novel design challenge is to develop ACEs that guide agents towards doing the right thing (morally speaking). Minimally, this presupposes an AED that will ensure that agent X achieves the kind of rationality, consistency, integrity and discrimination that we take to be essential for moral judgement.[9] To the extent that this involves the scrutinising of X's negative and positive attitudes, it is similar to the corrective technology that we imagine operating to support autonomy under conception (bb).

However, this conception of autonomy might be given any number of substantive moral articulations—for example, featuring utilitarian, human rights, or dignitarian principles (Brownsword 2008a). This being so, what further channelling influence might the technology appropriately exert? In Europe, at the beginning of the twenty-first century, where there is a political, legal and cultural commitment to human rights, it might seem appropriate that the technology should advise X as to the compatibility of X's free choices with respect for human rights. This is by no means a straightforward matter. Without attempting to be exhaustive, some of the more pressing and recurring questions to be addressed by agents who are committed to human rights are the following.

First, there is a large cluster of questions concerning which rights (negative and positive) are to be recognised and what the scope of particular rights is.[10] *Second*, there are questions arising from conflicts between rights as well as from competition between rights-holders.[11] Sometimes the conflict might be between one kind of right and another—for example, between the right to privacy and the right to freedom of expression. At other times, there are competing rights—that is, cases where two rights-holders present with the same general right. *Third*, because consent is an extremely important dynamic in a community of rights, the community needs to debate the terms on which a supposed 'consent' will be recognised as valid and effective (Brownsword 2004, Beyleveld and Brownsword 2007). *Fourth*, there are questions about whether there are limits to the transformative effect of the reception of rights (Beyleveld and Pattinson 2002). In particular, this invites reflection on the relationship between one community of rights and another (Brownsword 2007, Pogge 2008). And, *finally*, there is the vexed question of who has rights.[12] Do young children, fetuses, or embryos have rights? What about the mentally incompetent or the senile? And, then, what about non-human higher animals, smart robots, and, in some future world, hybrids and chimeras of various kinds?

Even though the members of a community of rights have a shared moral outlook, there is still plenty to debate and one wonders how far the technology could, or should, be designed to guide agents towards autonomous action relative to conception (cc). As with conception (bb), the more that we excavate conception (cc), the more problematic the idea of autonomy becomes and the less plausible it becomes to suppose that AED can do the trick. Once again, the best practical response might be to default to designs that are purely advisory, and what is more to advice that is tendered in a tentative way.

What are the implications of agents delegating their decision-making to smart computing aides?

Thus far, the possibility of X delegating decision-making to the technology has not been considered. However, relative to each conception of autonomy, would delegation salvage or sacrifice autonomy? Before we take up this question, we need to distinguish between different degrees of delegation.

Degrees of delegation

Suppose that agent X is equipped with a personal assistant (PA) that, inter alia, can search the Web for on-line deals that match X's requirements. If the PA senses that X needs a holiday and books one for him, this raises obvious questions about X's autonomy and responsibility for the booking. However, suppose that X has decided that it is time to take a holiday (possibly with a prompt from the PA) and now delegates some responsibility to the PA. Relative to the objective of booking a holiday, the delegation might be (i) advisory or executory, and (ii) closed or open. The permutations are as follows:

(A) advisory/closed (e.g., X instructs PA to identify, but not book, the best value 7-day beach holiday at a Mediterranean resort, with 4 star hotel accommodation, that costs in the range of 700–800 euros);
(B) executory/closed (e.g., X instructs PA to identify and to book the best value 7-day beach holiday at a Mediterranean resort, with 4 star hotel accommodation, that costs in the range of 700–800 euros);
(C) advisory/open (e.g., X instructs PA to identify, but not book, the best value holiday option for him); and
(D) executory/open (e.g., X instructs PA to identify and to book the best value holiday option for him).

In each case, we assume that PA acts on these instructions knowing X's preferences and without exceeding the terms of the delegation. In cases A and C, PA simply reports back to X (PA's function is purely advisory); in cases B and D, PA acts within the scope of its delegated authority by identifying and booking the holiday (PA's function is executory).

Delegation and autonomy

In each of the four examples above, what is the significance of X's delegation relative to X's autonomy? Given the three conceptions of autonomy identified above, and the four types of delegation, there are 12 cases to be analysed. Broadly speaking, and without trailing through each of the dozen cases, the results of this analysis are as follows:

First, if we operate with autonomy under conception (aa), cases A and C are entirely unproblematic because X has only partially delegated the task to PA; PA's role is advisory only. The effect of the delegation is to leave a distance between PA's recommendation and the act of booking. If X books a holiday in accordance with the information supplied by PA, we can still treat X's act as autonomous.

Second, if we operate with autonomy under conception (aa), cases B and D are less clear. There is no distance between PA's identification of the holiday and its booking on X's behalf (PA acts in an executory capacity). However, if X has delegated executive competence to PA in a way that is alert to the implications of such an authorisation, then this is arguably consistent with autonomy. The key act is not the booking but the delegation to PA. Provided that the delegation is made freely (in the sense required by conception (aa)), the booking nestles within the delegation, which is the governing autonomous act.

Third, if we operate with autonomy under conception (bb), and if we assume that PA correctly interprets X's interests, then cases A and C are unproblematic because, once again, X has only partially delegated the task to PA. If PA corrects for some inconsistency in X's preferences (possibly by rejecting a beach holiday because X is predisposed to skin cancer, or recommending a cheaper holiday because of X's limited funds), X has the chance to switch or confirm preferences before any decisive action is taken.

Fourth, if we operate with autonomy under conception (bb), cases B and D are unproblematic to the extent that PA books in line with X's considered and consistent preferences, or X's real interests. If this involves a deviation from the apparent terms of the delegation to PA (for example, if PA judges that a beach holiday is not in X's real interests), we need a more sophisticated interpretation of X's intention and his meaning in delegating to PA.

Finally, if we operate with a moral conception of autonomy (as in conception (cc)), the compatibility of PA's actions with X's autonomy will depend on the moral status of those actions—which, in turn, will depend upon which particular substantive moral criteria are treated as governing. In cases A and C, where PA simply gathers and evaluates information about X's holiday options, it is hard to see anything inconsistent with X's moral autonomy. In cases B and D, if we assume that the booking is within the scope of PA's authority, there might nevertheless be some immorality in the particular booking (e.g., because of the environmental costs or the exploitation of local labour) and, if the technology were morally sensitive, it would not book a holiday unless it were ethically clean.

What should we make of this? First, where the technology simply acts in an advisory capacity, there seems to be no direct challenge to X's autonomy (and, anyway, if X has self-consciously delegated this task to the technology, the controlling choice is still autonomous in sense (aa)). Second, where the technology acts in an executive capacity (casting the vote, booking the holiday, and so on), this raises a question about X's autonomy. However, provided that such acts have been authorised by X, we can still argue that X acts autonomously—in a sense, X's act of delegation keeps the technology at a distance. To be sure, where the decisions are made by technology that is routinely embedded in the surrounding environment (rather than by digital assistants or voting aides) the act of delegation might be less explicit; and, where the supposed authorisation is implicit, we will run into difficulties.[13] Finally, where we are operating with conceptions (bb) or (cc), agent autonomy is a more complex beast. The focus of concern here is the nature of the purposes chosen by X and the fact that X has delegated a competence to the technology cannot fully cover those concerns.

What is left of agent responsibility in an environment of autonomic computing?

Legal and moral discourse is underpinned by a set of assumptions concerning the practical rationality, the freedom, and the responsibility of agents. It is assumed that agents are capable of acting on normative signals, that they are capable of acting on the prudential and moral reasons that are given, and that they are rightly held responsible for non-compliance (Morse 2002). Having said that, whereas legal practice distinguishes between responsibility that implies blameworthiness (as is characteristic of the criminal law) and responsibility that implies no blame but simply the duty to compensate (as is characteristic of, say, product liability regimes), moral practice tends to focus on responsibility as blameworthiness (Cane 2002).

Where responsibility as blameworthiness is involved, we think it fair to hold an agent to account only if certain conditions are satisfied. One such condition is that the act in question truly was the agent's own act, that the agent acted with a sufficient degree of independence. In other words, one of the conditions of holding agents responsible for particular acts is that those acts are autonomous, that they are not the product of coercion, or duress, or hypnosis, or the like. We might also make this a condition of attributing responsibility in the compensatory sense but it would tend to be a less pressing consideration. At all events, for present purposes, the question is whether, where we require a degree of independence before we attribute responsibility, the development of ACEs will impact on our responsibility practice.

In the light of the analysis above, we can classify autonomic computing technologies as: (i) autonomy enhancing; (ii) autonomy eroding; and (iii) autonomy eliminating. Where the technology is autonomy enhancing, in the sense that it presents the agent with information that improves the agent's chances of making an informed choice, this seems like a clear case for attributing responsibility to the agent (in relation to the choice actually made by the agent). To this extent, autonomic computing might serve to strengthen the conditions for attributing responsibility. Conversely, where the technology is autonomy eliminating, and assuming that there has been no delegation of function, it would be out of line with current practices to hold the agent responsible. For example, if X's PA senses that it is time for X to take a break and books a holiday for X, it is unclear why X should be held responsible (if only as an insurer) for this booking—or, say, for an on-line fraud carried out by PA without X's knowledge or authorisation. The more difficult case—one into which we must also factor the complication of delegation—is that in which the technology is autonomy eroding; and this is the case to which we now turn.

Responsibility and autonomy-eroding technologies

Where an ACE steers or channels an agent towards a particular decision or act, but without actually compelling that act, it is arguable that the agent, having the final say, acts autonomously and remains responsible for the decision or act taken. Even where the environment gives the agent a strong 'nudge' towards a particular act, it is arguable that the agent's autonomy is respected—at any rate, provided that there is still sufficient distance between the agent and the environment (Thaler and Sunstein 2008). On this view, responsibility survives.

Against this view, we might argue that autonomy-eroding technologies exert an undue influence to the point that the agent fails to make an independent judgement before deciding or acting. It is no answer to insist that the technology is benign, that it is pushing the agent towards acts that are in the agent's own best interest, or the like. Without genuinely independent assessment, there is no autonomous decision (on pretty much any conception of autonomy) and there can be no responsibility.

It might be objected that, in themselves, technologies are not autonomy eroding. They only assume this character when the social context conduces to a swift adoption of the PA's recommendation. No doubt, this is correct and, indeed, it is a phenomenon that we see in many technology-free contexts where the culture encourages habitual adoption of the views or advice of influential others. But, what follows from this?

If we cannot draw a clear line between those cases of influence that we think are problematic relative to the attribution of responsibility and those that are not, we need to build in safeguards. For example, in the jurisprudence of English contract law, where one member of a family might exercise undue influence over another (typically, a husband exerting undue influence over his wife), financial institutions should ensure that the latter receives independent advice before entering into a potentially disadvantageous transaction.[14] The fact that agent X receives independent advice does not guarantee that X acts independently, but it suffices to treat X as responsible for the decision eventually made. In the same way, we might want to design in safeguards where an ACE has a tendency to erode autonomy.

Responsibility and delegation

Delegation, as we have seen, can be of different degrees. Where the delegation by X to PA is partial so that information is gathered (by PA) but no action is taken (by X) on the information, there is unlikely to be any question of X being responsible—unless the PA's information gathering is itself problematic (e.g., by infringing privacy or data protection codes). At any rate, it is where delegation involves PA acting in an executive capacity for X that we are likely to run into questions of X's responsibility. Taking a standard jurisprudential approach, we would say that, if X authorises Y to act as agent for X in some matter, then to the extent that Y acts within the scope of his authority, X will be treated as responsible for Y's acts (on X's behalf). Accordingly, where delegation by X to X's PA, or whatever, corresponds to X's authorisation of Y (as an agent), X remains responsible for PA's authorised acts.

Thus far, we have assumed that PA acts within the terms of X's express authorisation. But, what if the authorisation/delegation is merely implicit; or what if PA acts beyond the terms of the authorisation? The analogues in law are those of implied authority and of agents who exceed their authority. We now run into hard cases of various kinds, depending (with regard to *ultra vires* issues) on whether the party who deals with PA is aware, or might reasonably have discovered, that PA is exceeding his authority, and so on. We also have the possibility here of the producers of PA being held responsible (as insurers) for the injuries caused by a defective product (Wright et al. 2008).

Having said all of this, there is nothing in these difficulties that fundamentally challenges the idea that we can fairly hold an agent responsible for the actions that it freely chooses to take. Provided, as de Mul and van den Berg argue, the agent still holds the remote, or provided that the agent has explicitly delegated the use of the remote to a PA, it still seems appropriate to continue with our practices of blaming and shaming. It follows, though, that our difficulties seem to be set to intensify as ACEs become the rule rather than the exception; for, with pervasive ACEs, there might no longer be anything akin to a remote for agents to hold onto and, thereby, to remain in control.

The prospects for agents in a community of rights

So much for the mapping of the autonomy of agents in relation to ACEs. We can consider now how the development of ACEs would be viewed in a community of rights (that is, an ideal-typical community of the kind presupposed by modern European societies with their commitments to respect for human rights). Having sketched the characteristics of such a community, I identify three boundary-limiting considerations.

A community of rights: a thumbnail sketch

A community of rights is a particular kind of *moral* community. Hence, it must systematically embed a *moral* standpoint (in the formal sense); and because it is a community of *rights*, the governing substantive morality is rights-led (rather than utility-maximising or duty-driven). These defining characteristics call for some short elaboration.

First, as a *moral* community, the community of rights will hold its commitments sincerely and in good faith, it will treat its standards as categorically binding and universalisable, and there will be an integrity and coherence about its commitments as a whole. *Second*, as a community of *rights*, the acid test of moral action is to respect the rights of others, not to maximise utility or to act in accordance with the dignitarian duties that have become so influential in modern bioethics (Brownsword 2003, 2006). *Third*, a community of rights should be conceived as a society that views itself as a process rather than a finished product, constantly keeping under review the question of whether the current interpretation of its commitments is the best interpretation. *Fourth*, in a community of rights, the discourses of ethics and of regulation are regarded as both contiguous and continuous. It is not enough that regulation is effective and fit for purpose; the first priority is that regulators should have the right purposes (rights-respecting purposes) and that the regulatory standards that are set are legitimate relative to the community's rights values. And, *fifth*, in a community of rights, a will (or choice) theory of rights, rather than an interest theory of rights, is adopted; and, as a corollary, the paradigmatic bearer of rights is one who has the developed capacity for exercising whatever rights are held, including making choices about whether to give or to refuse consent in relation to the rights that are held.

Finally, it is worth repeating that, even with these shared characteristics, there is considerable margin for each community of rights to express and articulate its commitments in its own way—for example, as already indicated, from one community to another, there might be different views about the status of non-paradigmatic rights-holders, and especially so in relation to the details of the array of recognised rights (compare Beyleveld and Brownsword 2006).

Three key concerns in a community of rights

In a community of rights, the State will exercise a stewardship responsibility for the boundary-limiting conditions that are essential to the continuing viability of the community (Brownsword 2009a, 2009b). If an ACE is judged to compromise these conditions, it will be ruled out as contrary to the public interest.

(i) The conditions for moral community

In an ideal-typical moral community, agents freely choose to do the right thing and they do so for the right reason. Where the context for action is dense with the technologies of surveillance and recognition, there is a danger that agents comply with public standards for prudential rather than moral reason—as James Griffin has put it, '[t]here is no dignity in mere submission to authority' (Griffin 2008: 26). Given such a (human rights) perspective, the fundamental objection to panopticon surveillance is that it compromises the possibility for autonomous action—to be sure, prudential action can be autonomous action (assuming conceptions (aa) and (bb)), but there is no dignity in prudent compliance when the threat of detection and correction is overwhelming, and when there is no real opportunity to exercise one's conscience.

Where the context embeds the technologies of profiling and prevention, the degree of control and channelling might be even more exacting: in such an environment, it might be impossible for agents to act otherwise than they do (that is, to act otherwise than in accordance with the desired regulatory pattern of conduct). This altogether eliminates any plausible account of agents freely choosing to do what they actually do. It no longer makes any sense to hold agents responsible (Brownsword 2008b).

Having said this, should we eschew technological interventions that are designed to improve the safety of agents, particularly to safeguard agents against unintentionally harmful acts?[15] For example, following the inquiry, chaired by Lord Cullen, into the Ladbroke Grove train disaster (on the approach to Paddington station in London), the rail regulators sought to improve safety by introducing technologies that made it more difficult for drivers to miss signals that are on red or to drive past them regardless.[16] Similarly, in clinical settings, we might try to improve patient safety by designing medical equipment in ways that reduce the risk of misuse or misapplication. Primarily, such adjustments to the regulatory environment—the environment in which pilots and train drivers, surgeons and anaesthetists work—are designed to reduce the risk of harm being *unintentionally* caused to passengers and patients. Although such measures are targeted at acts that are inadvertent, or careless, the effect of such technological fixes is that the regulatory environment is adjusted in a way that impacts on potentially harmful acts, whether unintentional or intentional. Does this matter?

Surely it is implausible to argue that technological interventions that are designed for safety should *not* be adopted because they will give, say, train drivers or pilots less opportunity intentionally to inflict injury on their passengers. Unless train driving or aircraft flying offers a unique opportunity for the cultivation of moral virtue, there seems little sense in putting the lives of passengers at risk simply so that pilots and train drivers have the opportunity to do the right thing. If a community is short of such opportunities, there must be better places for their creation than on the railways tracks or in the skies. Does it follow, then, that where regulators employ technological design in order to minimise the risk of unintentional harm there is never a problem? The short answer is that it does not. It is important to consider just how the technological innovation changes the regulatory environment. For example, if surveillance cameras were to be introduced into clinical settings, including operating theatres, this would impact on both clinicians and patients; and, where a particular design impinges on a previously discretionary band of application (for example, concerning 'single use' medical devices), there might be some negative consequences. This is far from suggesting that technological measures of this kind should be eschewed—the point is the more modest one that, even where the regulators are targeting unintentionally harmful acts, we need to check whether regulatory effectiveness is being procured at some cost in regulatory legitimacy.

(ii) The agency commons

Famously, Alan Gewirth (1978) has argued that the logic of agency is such that one must have (at minimum) a positive attitude towards the generic conditions of agency and, concomitantly, a negative attitude towards unwilled interferences with those conditions. When Gewirth specifies the generic conditions of agency, he does so in terms of an agent's freedom and well-being. So long as we think of this as supplying the substance of the generic conditions, it is tempting to distinguish between *my* generic conditions (my freedom and well-being) and *your* generic conditions (your freedom and well-being). This invites deep resistance to the argument as it moves on; and, even though such resistance is tractable,[17] it can be eased if the significance of the essential *context* for agency is drawn out. To be sure, some threats to agency arise from the isolated aggressive acts of our fellow humans but they also arise from acts that are routinely damaging to the environment, to public health, and to a sense of security. In polluted environments, agents die; without clean air and water, agents perish; if exposed to epidemics and pandemics, agents fall ill and sometimes die; and, in settings that are fraught with danger and insecurity, agents operate in a defensive mode. If agents are to flourish, they need clean air, clean water and a context that does not elicit defensive measures. This context is, so to speak, the agency commons (Brownsword 2009b), a commons that each and every agent must view in a positive light.[18]

It follows that, in a community of rights, there will be intense concern about any technological application that compromises the agency commons. The kind of technologies that are at issue in this paper are unlikely to threaten the health, safety, and well-being of agents. Rather, the threat is to a cultural context that encourages agents to make their own life choices. And, more particularly, where the choices that matter are moral choices, the threat is to a cultural context that encourages agents to make their own moral judgements.

(iii) The constitutionally protected scheme of rights

The third deep concern of a community of rights will be that ACEs do not infringe constitutionally protected rights. Some of these rights, particularly the rights of agents to their freedom and basic physical and psychological well-being, intersect with the agency commons. However, if we think of the agency commons as representing the essential setting or staging for agency, then it is the constitutional protections that apply to the actions of agents within that setting. In other words, if the agency commons provides a setting that is conducive to the freedom and well-being of agents, it is the constitutional rights that protect agents who interact and transact on the commons.

Because ACEs go to the very setting for agency, it will be the protection of the commons that is the focus of concern in a community of rights. Even so, there will also be concerns about the way in which ACEs or digital assistants might impinge on the protected rights of agents. For example, there might be concerns about the injury and damage that can be caused where the technology of ambient environments fails. Anticipating such failures, the community will consider precautionary measures. However, as with most of the other matters discussed in this paper, there are complex calculations to be made before the community can arrive at a sensible judgement about the adoption of such measures (Beyleveld and Brownsword 2009).

Conclusion

Any attempt to map the implications of ACEs in relation to our understanding of agency and autonomy and our practices of attributing responsibility, soon runs into difficulty. On one axis, we do not have a standard conception of autonomy (in this paper, I have identified three quite different conceptions); agents can delegate their tasks with varying degrees of specificity and completeness; and there is more than one sense of responsibility in play. On the other axis, there are many technological designs that can serve to eliminate, to erode, or to enhance the things that we value. If we are to identify all permutations, we face an enormously complex task—a task probably better left to powerful computing technologies rather than less competent human agents.

How, then, should we proceed? In my view, the fundamental challenge is to stabilise our ideas of autonomy and responsibility. To do this, we need to be clear about what it is that we value. I have argued elsewhere that we have good reasons for believing that a community of rights is the ideal-type to which, as (rational) agents, we should aspire (Brownsword 2008a). In such a community, it is plausible to suppose that the voluntariness element of autonomous action would be understood in very much the terms outlined by de Mul and van den Berg, and responsibility would follow suit. Accordingly, it would be important, either through the design of the technology or by dint of the cultural understanding or both, to ensure that agents are constantly reminded that, even in ACEs, the final decision is theirs to make.[19]

Stated shortly, in a community of rights, there is a triple bottom line, namely: that ACEs (like any technology) should be compatible with (i) the essential conditions of moral community, (ii) the agency commons, and (iii) the array of recognised rights. Insofar as ACEs reduce the 'affordances' available to agents (Zittrain 2008: Ch. 4), especially by eliminating the possibilities for intentional (even supposedly immoral) action, the community needs to ask not simply whether this impinges on autonomy and responsibility but whether this runs counter to the community's aspiration that agents should freely make their best moral judgements and act on them.

Notes

1 An earlier version of this paper was given at a Panel on 'Autonomic Computing, Human Identity, and Legal Subjectivity', by way of a response to Jos de Mul and Bibi van den

Berg's paper, 'Remote Control'. The panel met under the auspices of the conference on Computers, Privacy and Data Protection, held in Brussels, 16–17 January 2009. I am most grateful to participants at the panel for their comments; but, of course, the usual disclaimers apply.

2 See, too, the various (not so futuristic) scenarios sketched in Brenner (2007: Ch. 1).

3 Compare the entry for 'Personal Autonomy' in the *Stanford Encyclopaedia of Philosophy* (2009), where four principal approaches are identified, namely: coherentist, reasons-responsive, responsiveness-to-reasoning, and incompatibilist.

4 This is closest to a coherentist account of autonomy, see note 2 above; and, generally, compare Dworkin (1988).

5 Compare Mireille Hildebrandt's distinction between 'regulative' and 'constitutive' technological features: see, Hildebrandt (2008).

6 The leading case in this somewhat unsettled area of English law is *Royal Bank of Scotland v Etridge (No 2)* [2001] UKHL 44.

7 Compare the sustained critique of the 'misuse' legislation in Brenner (2007).

8 But, is there any basis for this assumption? Where the vote is cast without any reference back to X, even if the vote that is actually cast is in line with X's real interests, it is hard to see how this represents a free act by X (in the sense proposed by de Mul and van den Berg)—unless, of course, X has freely delegated this executive function to the voting aide. See further below at 74.

9 For the distinction between morals in an 'anthropological' and 'discriminatory' sense, see Dworkin (1978: 248–53).

10 Compare, e.g., Griffin (2008: Ch 5) (on positive rights) and Chs 12 and 13 (on the scope of the particular rights to life and death and to privacy).

11 Again, compare, e.g., Griffin (2008: Ch 3) (which opens, at 57, with the insightful remark that there 'is no better test of an account of human rights than the plausibility of what it has to say about rights in conflict', and that there 'is no better way to force thought about human rights to a deeper level than to try to say something about how to resolve conflicts involving them').

12 Again, compare, e.g., Griffin (2008: Ch 4).

13 We can imagine cases where the delegation is based on a failure to opt out, or by virtue of constructive notice, and the like. In an analogous case, see the brilliant critique of the fictionalising of consent in Harmon (1990).

14 Again, the leading case in the modern law is *Royal Bank of Scotland v Etridge (No 2)* [2001] UKHL 44.

15 Here, I am drawing on Brownsword and Somsen (2009).

16 There is some discussion of 'automatic controls' in the Inquiry Report: see Health and Safety Commission (2001), paras 12.25–12.28; and for reliance on engineered solutions to 'signals passed at danger' see Medical Research Council (2008), 25 (Bob Chauhan's presentation).

17 The classic defence of the Gewirthian argument, through all its stages, is in Beyleveld (1991). For a recent exchange, see Chitty (2008); and Beyleveld and Bos (2009).

18 Compare the argument in Westra (2004).

19 However, by way of qualification, it should be said that where ACEs eliminate choice by excluding the possibility of unintentionally or carelessly causing harm to others (e.g., transport safety systems), this would not always be seen as objectionable, all things considered, in a community of rights.

References

Beyleveld, D. (1991) *The Dialectical Necessity of Morality: An Analysis and Defense of Alan Gewirth's Argument to the PGC*, Chicago: University of Chicago Press.

Beyleveld, D. and Bos, G. (2009) 'The Foundational Role of the Principle of Instrumental Reason in Gewirth's Argument for the Principle of Generic Consistency: A Response to Andrew Chitty,' *King's Law Journal* 20:1–20.

Beyleveld, D. and Brownsword, R. (2006) 'Principle, Proceduralism and Precaution in a Community of Rights,' *Ratio Juris* 19: 141–68.

—— (2007) *Consent in the Law*, Oxford: Hart.

—— (2009) 'Complex Technology, Complex Calculations: Uses and Abuses of Precautionary Reasoning in Law' in Marcus Duwell and Paul Sollie (eds), *Evaluating New Technologies: Methodological Problems for the Ethical Assessment of Technological Developments*, Dordrecht: Springer: 175–90.

Beyleveld, D. and Pattinson, S. (2002) 'Horizontal Applicability and Direct Effect,' *Law Quarterly Review* 118: 623–46.

Brenner, S. W. (2007) *Law in an Era of 'Smart' Technology*, New York: Oxford University Press.

Brownsword, R. (2003) 'Bioethics Today, Bioethics Tomorrow: Stem Cell Research and the "Dignitarian Alliance"' *University of Notre Dame Journal of Law, Ethics and Public Policy* 17: 15–51.

—— (2004) 'The Cult of Consent: Fixation and Fallacy,' *King's College Law Journal* 15: 223–51.

—— (2006) 'Cloning, Zoning and the Harm Principle' in S.A.M. McLean (ed), *First Do No Harm*, Aldershot: Ashgate: 527–42.

—— (2007) 'The Ancillary Care Responsibilities of Researchers: Reasonable but Not Great Expectations,' *Journal of Law, Medicine and Ethics* 35: 679–91

—— (2008a) *Rights, Regulation, and the Technological Revolution*, Oxford: Oxford University Press.

—— (2008b) 'Knowing Me, Knowing You—Profiling, Privacy and the Public Interest' in Mireille Hildebrandt and Serge Gutwirth (eds), *Profiling the European Citizen*, Dordrecht: Springer: 362–82.

—— (2009a) 'Rights, Responsibility and Stewardship: Beyond Consent' in H. Widdows and C. Mullen (eds), *The Governance of Genetic Information: Who Decides?* Cambridge: Cambridge University Press: 99–125.

—— (2009b) 'Friends, Romans, Countrymen: Is There a Universal Right to Identity?' *Law Innovation and Technology* 1: 223–49.

Brownsword, R. and Somsen, H. (2009) 'Law, Innovation and Technology: Before We Fast Forward—A Forum for Debate,' *Law Innovation and Technology* 1: 1–73.

Cane, P. (2002) *Responsibility in Law and Morality*, Oxford: Hart.

Chitty, A. (2008) 'Protagonist and Subject in Gewirth's Argument for Human Rights,' *King's Law Journal* 19: 1–26.

de Mul, J. and Berg, B. van den (2011) 'Remote Control' (in this volume).

Dworkin, G. (1988) *The Theory and Practice of Autonomy*, New York: Cambridge University Press.

Dworkin, R. M. (1978) *Taking Rights Seriously*, rev. ed., London: Duckworth.

EC Health and Consumers Directorate-General (2009) *Data Collection, Targeting and Profiling of Consumers for Commercial Purposes in Online Environments* (Brussels, 05.03.2009, SANCO/B2/GA/SR/GR).

Gewirth, A. (1978) *Reason and Morality*, Chicago: University of Chicago Press.

Griffin, J. (2008) *On Human Rights*, Oxford: Oxford University Press.

Harmon, L. (1990) 'Falling off the Vine: Legal Fictions and the Doctrine of Substituted Judgment,' *Yale Law Journal* 100: 1–71.

Health and Safety Commission (2001) *The Ladbroke Grove Rail Inquiry*, London: HMSO.

Hildebrandt, M. (2008) 'Legal and Technological Normativity: More (and Less) than Twin Sisters,' *TECHNE* (12) 3: 169–83.

Klang, M. (2006) *Disruptive Technology*, Göteborg: Göteborg University.

Medical Research Council (2008) *Report of the MRC/Wellcome Trust Workshop on Regulation and Biomedical Research* (held May 13–14, 2008), London.

Morse, S.J. (2002) 'Uncontrollable Urges and Irrational People,' *Virginia Law Review* 88: 1025–78.

Pogge, T. (2008) 'Aligned: Global Justice and Ecology' in Laura Westra, Klaus Bosselmann and Richard Westra (eds), *Reconciling Human Existence with Ecological Integrity*, London: Earthscan: 147–58.

Stanford Encyclopaedia of Philosophy, 'Personal Autonomy' (http://plato.stanford.edu/entries/personal-autonomy/index.html) (last accessed July 15, 2009).

Thaler R.H. and Sunstein, C.R. (2008) *Nudge: Improving Decisions about Health, Wealth, and Happiness*, New Haven: Yale University Press.

Westra, L. (2004) 'Environmental Rights and Human Rights: The Final Enclosure Movement' in Roger Brownsword (ed), *Human Rights: Global Governance and the Quest for Justice Vol IV*, Oxford: Hart: 107–19.

Wright, D., Gutwirth, S., Friedewald, M., Vildjiounaite, E., and Punie, Y. (eds) (2008) *Safeguards in a World of Ambient Intelligence*, Dordrecht: Springer.

Zittrain, J. (2008) *The Future of the Internet*, London: Penguin Books.

Chapter 5

Rethinking human identity in the age of autonomic computing

The philosophical idea of trace

Massimo Durante

Introduction: the autonomic computing system

As IBM has remarked,[1] we progressively enter into the age of 'autonomic computing', which has risen 'to the top of the IT agenda because of the immediate need to solve the skills shortage and the rapidly increasing size and complexity of the world's computing infrastructure'. What is thus autonomic computing? It is 'an approach to self-managed computing systems with a minimum of human interference. The term derives from the body's autonomic nervous system, which controls key functions without conscious awareness or involvement', whose purpose is 'to realize the promise of IT: increasing productivity while minimizing complexity for users', by designing and building 'computing systems capable of running themselves, adjusting to varying circumstances, and preparing their resources to handle most efficiently the workloads we put upon them'. This is worth noticing: autonomic computing displaces the role of human intervention while relying on an unpersuasive analogy with 'the body's autonomic nervous system'. This metaphor, however, has a precise role: it represents the subject as a part of a whole, which precedes the subject and runs after it. According to this metaphor, the subject emerges and is profiled only out of the interactions that are embedded into the self-managing system. We are, me and all the other objects, 'data' that are part of the same space (computing infrastructure). This is well recognised by the IBM proposal: 'This new paradigm shifts the fundamental definition of the technology age from one of computing, to one defined by data. Access to data from multiple, distributed sources, in addition to traditional centralized storage devices will allow users to transparently access information when and where they need it'.

It is not the only metaphor at play. The autonomic computing is also compared to 'the self-governance of social and economic systems': this metaphor is of great interest from a social and legal standpoint, and it directly concerns the role of human agency. On the one hand, as said, human intervention in the system is displaced: our physical presence is reduced to a 'human interference', while our networked digital presence stems only from the access to data from multiple, distributed sources. On the other, human intervention is kept alive and entrenched with the initial, limited role of selecting the general goals that the autonomic system is expected to achieve on behalf of us: 'While autonomic systems will assume much of the burden of system operation and integration, it will still be up to humans to provide those systems with policies – the goals and constraints that *govern* their action' [my emphasis] (Kephart and Chess 2003: 48).

By reading the Manifesto of autonomic computing and following the second metaphor at play (concerning self-governing systems), I can highlight three main potential problems (potential, since the autonomic computing is not yet fully deployed), which seem to me relevant from a social and legal standpoint.

The general/specific level of decisions

As many self-governing systems, in the autonomic computing system there are two levels of decisions. One level that is made of general decisions brought forth by human beings, which are entrenched with the role of selection and specification of general goals to be autonomously pursued by the system (Kephart and Chess 2003: 42–43). And another level that is made of specific, operational decisions, which have to implement the general decisions and obtain accordingly the goals specified: this is the level at which the autonomic system operates. In any self-governing system there is always a problem of communication or articulation between these two levels of decisions, as recognised in the Manifesto: 'The enormous leverage of autonomic systems will greatly reduce human errors but it will also greatly magnify the consequences of any error humans do in specifying goals' (Kephart and Chess 2003: 48). I will not focus my attention on the technological aspects of this problem affecting the design of the system, but, more particularly, on three sub-problems that are concerned with the metaphor of a self-governing system.

Prediction/anticipation

According to the vision of autonomic computing, human beings are entrenched with the role of determining and specifying the general goals pursued by the system. This requires them to have a full capacity of prediction, in order to enable the autonomic system to display its role of execution and anticipation. Human beings are thought of as absolute subjects provided with a whole self-consciousness and awareness of what are the desired goals to be stated. This view is well summarised by the idea of human consciousness as a *legislator* of all things. However, in the self-governance of complex socio-legal systems, we are acquainted with the fact that the legislator is no longer in the position of having all the information required in order to make, from the beginning, full predictions about the general goals to be pursued, which are only specified over time by means of multiple and distributed contributions brought about by intermediaries. This leads us to a second sub-problem, that of mediation.

Mediation

In the self-governance of complex socio-legal systems, there are always interme-diaries between these two levels (e.g., between the law and its observance: judges, legal scholars, public administrators, etc.). Intermediaries are entrusted with the role of monitoring how general goals are made defined and unambiguous and how they are specifically applied. According to more radical legal views, these interme-diaries are entrenched with the role itself of specifying these general goals. In both cases, they are meant to mediate between the two levels of decisions on behalf of who have taken the general decisions. This further specification of how general goals are to be concretely applied occurs normally within a public procedure (e.g., trial or administrative procedure) where all people concerned by the execution of general decisions participate. In that context personal responsibilities are called forth and recognised. This leads us to a third sub-problem, that of responsibility.

Responsibility

In the autonomic computing system, autonomic managers seem to be entrusted with this crucial role of intermediaries: autonomic managers are repeatedly asso-ciated with the term 'responsibility' (Kephart and Chess 2003: 44). It seems to me that this reference is again only metaphorical – even if I can understand that this metaphor is decisive for the social acceptance of the autonomic system. This term should be replaced, for instance, with a proper term, like that of 'task'. Being responsible means, from a legal and moral standpoint, something more than being entrenched with a particular task: it means being called upon to respond for some-thing to someone else before a third person (like a judge). This is precisely the situation in which human agency is judged and recognised, retrospectively, as autonomous and free. Autonomic managers' responsibility seems to be, at best, an assignation of tasks without any precise form of legal or moral accountability.

Privacy

In the information age our own identity is made of all the data referred to us (which give form to one's account). However, this process of referring data to one's account has never been an activity that the subject has achieved by herself. Our identity has always been made of data that come from multiple and distributed sources, and often from sources that are potentially conflicting with each other. From this point of view, the autonomic computing system does not differ too much from what happened in the past. At all times there is a plurality of different sources in referring personal data to one's account (i.e., in constructing an identity), and there is often a competition among the different sources. Autonomic computing is a powerful, distributed source in collecting and referring data to a personal account: for this reason, it could alter the conditions of this competition among different sources of data. As in many other cases, the competition has thus to be fair, in order to be socially accepted. Fairness is a fundamental value, according to which all the participants have to be given at least equal opportunities and means in the process of defining their own identity. The reference to appropriate 'measures' (Kephart and Chess 2003: 47) that are to be established, in order to protect personal data, do not seem to take into full account this need of fairness, which plays a crucial role as regards to privacy, since privacy in the information age is no longer to be understood as a mere protection of data from interference but, more radically, as a fair construction of personal identity (out of a competition among multiple, distributed sources of data).

Trust

The functioning of an autonomic computing system is achieved, at least to some extent, by means of software agents. This requires us to helping 'humans to build trust in autonomic systems', that is, in artificial autonomous agents. Contrary to the previous points (which I will expand in the present chapter), which show some criticism on the autonomic system as modelled upon a self-governing system, I agree that the requirement and development of trust in multiagent systems is crucial, since trust is the fundamental basis of every form of social cooperation, whatever be the environment in which such cooperation is meant to be deployed. In both computer and social sciences, it is more common to speak of mere reliance, with regard to artificial autonomous agents, than of trust that is reserved to human interactions. However, a growing number of scholars, by drawing from different perspectives, is going to put into doubt such assumption and to develop a revised notion of trust that applies also to software autonomous agents, since trust does not necessarily require a human touch and a relation between human agents (Durante 2008, Ghanea-Hercock 2007, Grodzinsky at al. 2009, Taddeo 2008, 2009). I cannot expand this point here, which I have discussed elsewhere (Durante 2010).

In the present chapter I would like to concentrate my attention upon the philosophical premises that underlie a self-governing legal-social system, which is assumed, at least metaphorically, as a model for the autonomic system. According to the points highlighted, there is a clear distinction between what humans are entrusted to do (to select general goals), and what the autonomic system is entrusted to do (to achieve the established goals). This implies the following: humans are capable of fully anticipating the general policies and goals of the system and, accordingly, of designing how to achieve them without their own interference. Humans are *legislators* who can determine the reality since they have already determined who they are (their identities, expectations, desires, policies, goals, etc.). There would be a perfect, direct correlation between human beings' identity and intentionality (*what is determining*) on the one hand and reality (*what has to be determined*) on the other. The autonomic vision seems to reaffirm a model of self-governance based on full human autonomy, whilst it suggests exactly the opposite: a system displacing as much as possible human intervention. The model of self-governing system based on the distinction between who selects goals (the absolute subjects) and who achieve them is problematic. The subject is always already implied in the situation that she attempts to govern, and her intentionality stems from this relation of mutual implication with the others that are part of the environment she is plugged in.

Correlation vs. mutual implication

As said, the vision of autonomic computing does not confront us with something totally new from a theoretical standpoint. This requires us to realise what are the conditions of human interaction in multiagent systems and hence what is the nature of the agents we interact with. How do agents behave in the cyberspace? How can we conceive their digital presence in the net? How can we capture their identity, decisions or actions, and assign responsibility for them? This demands us to reconsider the notion of identity based on the distinction between the subject (the 'I') and the object (the 'data'). How would it be still possible to refer to an 'I', when this has become 'part' of a self-managing system of data? How to recognise and protect our personal identity and autonomy, if it is no longer possible to make reference to an 'I'?

We have progressively accepted the idea that the subject is no longer self-projecting and transparent (i.e., the philosophical premises of the absolute subject). The subject is not *always already* constituted in her own prerogatives: she does not simply govern reality as a legislator. The subject is constituted in response to the claims of the others and the challenges of the environment that she is constantly confronted with. The subject cannot be conceived as someone that is precedent to the system and can fully design it thanks to this precedence: the subject is always embedded in the system. If understood in this way, the digital 'presence' of the subject does not disappear but it has to be rethought. How we can say that we are 'present' in the net in our personal identity and with our autonomy? If access to data becomes the norm that rules the system, how would it be possible to have a private self to safeguard, which is not already made 'present' within the net? How can we protect privacy that is nowadays not only a mean of protecting our personal data but, first of all, of constructing our identity?

The convergence between offline and online reality and the interplay between the human and the autonomic management of data cooperate to question the representation of space and of the subject that prevails in the modern age: this representation is based on the relation inside/outside, that is, on the depth between an inner and an outer side. This spatial metaphor governs also the difference between the subject and the object and is the philosophical premise of a subject already constituted in all her prerogatives and situated at the centre of the reality. In this framework, autonomy is viewed as a form of intentionality explained by a model of perfect adequacy between the subject and the object: autonomy would be the prerogative of a subject capable of determining the reality according to her plans (Serres 2009). The idea of a *correlation* is the epistemic guarantee of the direct determination of reality by the absolute subject. Because of the correlation, such determination would be the sign of our autonomy. We oppose to this model the idea of a *mutual implication* between subject and reality, according to which an embodied subject modifies reality and is modified by it. This means that the subject is part of the modified reality and hence that self-transformation is an inherent characteristic of subjectivity. The traces of our determinations cannot explain the reality as the direct representation of our will and, correspondently, of reality: the trace has to be transformed into a sign, in order to explain the reality.

The autonomic system seems to be based on a model of correlation: humans are conceived as legislators selecting goals and constraints according to which the reality of the system should be configured automatically and autonomously, that is, without human intervention in the system. The role of humans is both privileged and displaced. According to this model, human decisions are the 'sign' of how reality will be configured without any further human interference. This model is not persuading. Human goals are only the traces of how reality will be constituted by means of human intervention, which is needed at least in a social and legal self-governing system.

The model of *mutual implication* between subject and reality will be analysed through the reference to the relation between traces and signs for two reasons: the inscription of traces is the way we have been using from the ancient times to manage the surplus of data, to keep the memory of what happens (Ferraris 2009: 225 et ss).[2] Furthermore, the trace represents philosophically what can be transformed into a sign, without ceasing however to be a trace.

The model of correlation has been first put into discussion in the analytics of existence of Martin Heidegger (2008) and in the ethics of Emmanuel Lévinas (1996: 64),[3] from diverse perspectives. Heidegger introduces an ontological perspective, constructing the notion of *Dasein* as a relation of mutual implication between the ontological and the ontic dimension of Being. Lévinas introduces an ethical perspective, conceptualising the idea of subjectivity as a relation of mutual implication between the finite dimension of human existence and the infinite dimension of responsibility towards the others. Derrida has constructed upon the relation of mutual implication a profound interpretation of the role of the trace, on which is based his notion of difference (Derrida 1967, 1972). Maurizio Ferraris has recently given a full account of the ontological and epistemological status of the trace that bridges continental and analytical philosophy and offers a deep understanding of the documental dimension of social reality in the information age (Ferraris 2007, 2009).

In this perspective, the model of correlation not only seems to be incapable of explaining the construction of human agency and identity but also appears unable to account for the change in the representation of space and of the subject in the information society. The main characteristic of space and of the subject is their documentality (Floridi 2003),[4] it being made of documents, data or traces, of what can be inscribed on a support, the trace being the point of reference of what can be retrieved and identified.

The idea of mutual implication not only is in a better position to explain reality but is also more able to account for the reshape of space and subject in the information society. Notably, the subject is no longer viewed as the author directly determining the object and governing reality (the legislator). She is part of the documental fabric of reality in which she is implied. The subject is no longer understood as the external author of the traces left, but as part of the process of leaving and retrieving traces. The 'author' (and thus the authority or the governmentality, to employ the term that Ferraris borrows from Foucault)[5] grows out of the process that transforms the traces, left and trailed, into signs: this process necessarily requires the presence of a 'third' (due to the 'absence' of the author) that is charged with the reflexive role of reconstructing a history throughout the traces left. The necessary role of a third pole excludes from the model of mutual implication every form of dichotomy, both the idealistic dichotomy between subject and object and the hermeneutical dichotomy between author and text.

The philosophical significance of traceability

We aim to show that the philosophical idea of trace/sign implies a relation of mutual implication between the subject and reality. I will make use of Lévinas' distinction between the trace and the sign. According to Lévinas, the inscription of the trace is the insertion of space within time, the point where the world inclines itself towards the past and a time (Lévinas 1996: 67). It is the trace that determines the insertion of the dimension of space within that of time: this dimension of space cannot be understood as a common space, and the time of the trace is a diachronic time. The trace is the trace of what is no longer there: it is the trace of what is somewhere else. It never indicates as such a full 'presence' (even if differed) but only an absence: it is not the mark of an already established identity. In this sense, we can say that a trace, a mark, is *left*. This means that what leaves the trace is lost forever from the regime of presence: it is substituted. Yet, the substitution of the trace is indirect, since it is only the mark of what has been lost and cannot, as such, be retrieved (Durante 2007a).

The trace is not a link bringing us back to something or someone that we could localise elsewhere. It does not deliver any representation, as the sign does. However, since the trace is etched or inscribed, it possesses some materiality, which makes the trace an *entity*, contrary to a sign, which only has a *status*. The mode of being of a sign consists, as is known, of making a reference to something else, in showing what is not present (representation of the absent): this confers a status to the sign but it does not provide for the materiality of an entity. The sign substitutes what is absent and shows it: this means that the sign introduces what substitutes in the dimension of what is public[6] and general. This substitution is the domain of publicity and generality, because the content of a sign can be shared as a common reference. The publicity and generality of human goals require the transformation of traces into signs according to this regime of substitution.

However, the substitution of the trace is not by itself public. It indicates that we have lost something, we have been *deprived* of something: it is the regime of what is *private*. Even if the regime of the trace is different from that of the sign, they are not opposed. The relation between trace and sign is no longer conceived of as a mere opposition, but like a mutual implication. The trace can be identified and taken as a sign (Lévinas 1996: 69). This identification does not eliminate the material dimension of the trace (the inscription on a physical or digital support). The *inscription* of the trace assures the possibility of iteration and hence of justice: it makes adjudication possible, as the trace can become the content of evidences, being identified *as a sign*.

The space opened by the trace can be delimited and localised by the sign. This space becomes an enclosed, identified space: the space of identification. We move from the *commonly shared reference to* a trace (reference to a material inscription) to the *common shared reference as* a sign (reference to an interpreted shared content). In order to identify something or somebody, we need to determine this space of identification, which is the space of the third, because the author is absent and we need a third term assuring the transformation of the trace into a sign. In this sense, the third (this process of transformation or act of mediation)[7] becomes the 'author' and hence 'the authority' governing the identification of signs (general goals). In order to identify a trace and transform it into a sign, it is necessary to inscribe a new trace, which identifies the former and constitutes it as a common reference. The inscription of a second trace sets the space of identification, which is the 'transparent' space of our social and legal identity. We understand why in the idea of the autonomic system transparency is underlined.[8] However, transparency is not a first-order property but only the result of this second inscription of trace: transparency is always created. At the same time, the inscription of a new trace opens as well a new (*virtual*)[9] space, not yet identified. This process is not finite and exists over time: the trace can be endlessly re-identified and constituted into diverse signs.

In this perspective, the trace is a *point of reference*, that is, a point to which we can refer before this point is constituted as a common reference. Thanks to its materiality (traceability, in the digital world), the trace is the point of reference that allows us to relate reality with an identified meaning. The space of identification is, simultaneously, the space where the substitution operates (the space of what is *public and general*) and, at its limits, the space where the substitution is no longer possible (the space of what is *private and individual*).

In this sense, the relation between the trace and the sign is not expressed by means of opposition but of mutual implication. It follows from that that there is no ontological distinction in principle between what is private and what is public: there cannot be ontologically private data and ontologically public data. Whatever is private can be turned into public and vice versa (Mathieu 2004: 34). Such a distinction concerns two possible modes of being of the same entity: the trace. The space of privacy is already part of this mutual relation of implication and, for this reason, it is not part of the opposition between an inside and an outside (of the system). The space of privacy is the space of what has been inscribed. However, since the trace is inscribed, the individual space of privacy becomes a space where the trace is exposed to possible identification and generalisation. The trace allows people to enter into mutual relation: this transformation of the trace into a common sign creates a 'community' that presupposes the human mutual relation. This means that human intervention cannot be limited to the setting of general goals, which become general only after a necessary mediation and a mutual relation. What is more, not everything can be substituted: there is a sense in which our personal identity cannot be substituted (transformed into a public and general sign) but remains private and individual.

The space of privacy and the problem of substitution

The space of privacy is not constituted by a determined content (i.e., by a set of data). On the contrary, as suggested through the example of traceability, it is constituted by the power to refer something (a set of data) to a point of reference. Vittorio Mathieu interprets the *self* as being this point of reference (which we have merely understood so far as a trace):

> What distinguishes the public from the private is not thus a different content, but a different reference of the same content: the private is referred to a point, to which there is no reference in the public. [. . .] It is this reference to the self as a 'point' that grounds privacy and not the simple delimitation of a domain more restricted within a wider space.
>
> (Mathieu 2004: 39)

This means that the protection of privacy is not to be understood as the protection of a delimited space to be interdicted. On the contrary, the defence of privacy requires a human capacity to act (Mathieu 2004: 33), that is, the capability of a person to refer something to the self, the power of self-identification, which is an individual and a relational activity. The private and the public worlds, ontologically coextensive, do not exist by themselves but are the result of a human activity, which never consists of isolating the individuals from the society but in positing them within it.

At present, we can elucidate our idea: the reference to the self is nothing else than the reference to a trace, to a 'point of reference'. Whatever is the candidate for this point of reference ('self', 'body' or 'person'), it does not make a difference, because all these references are only traces of the (legal and philosophical) need to have a point of reference to be trailed and identified. Furthermore, through the commonly shared reference to a trace, we can construct the distinction between what is personal and what is impersonal. We never do it by ourselves, since this reference must be shared. This reference entails a society and a bundle of human relations: this implies the *power* to trace a difference between personal and impersonal but also the *lack of power* shown by the articulation between trace and sign. The trace can be always taken as a sign: this is the risk of alteration on which postmodernism has focused its attention. However, the trace can never be totally reduced to its evidences, because its transformation demands the inscription of new traces: this is the temporal *surplus* of the trace over the sign.

To protect privacy, that is, to assure identity, is to protect this power of self-identification, which always exists within the relation of mutual implication between the privateness of trace and the publicness of sign. The power of self-identification is associated with the power of being identified by someone else: self-identification cannot be defended as a prerogative of an *absolute subject* (overlooked in the post-modern age, but still implied in the vision of autonomic computing, if the role of humans is to determine general goals without any further intervention), but only as part of a *communication* with the power of being identified by the others. This brings us back to another point raised by the autonomic system. Even if this system might raise problems for personal privacy and individual autonomy because of its anticipatory managing of data, it confronts us with a serious, philosophical question about human identity. The informational representation of human identity consists of a set 'data from multiple, distributed sources, in addition to traditional centralized storage devices'. This proposition not only is as a depiction of an autonomic system but as a description of our culturally embedded personal identity.

In this perspective, we have to recall one point that is crucial and marks a possible criticism upon the autonomic system *as modelled on a self-governing system*: in every system, humans are not only concerned with relations in which they are substitutable, but also with relations in which they are not. According to the autonomic system, humans are not substitutable only in relation to the selection of general goals. However, they become afterwards immediately substitutable thanks to the generality of their decisions. What does this generality philosophically imply? The data constituting our identity can also be referred to someone else. We are often substitutable also in a self-governing system. However, the autonomic system of data seems to make us infinitely more substitutable: it widens the domain of substitution. Furthermore, it is not substitution that constitutes our personal identity: *we are who we are in what we cannot be substituted for*. More important, it is in what we cannot be substituted for that we are free (autonomous). This strategy accomplished by Heidegger is crucial: the idea of insubstitutability is a way of thinking of *individuation* no longer by means of a quality but throughout a relation where the subject cannot be substituted.

We should focus on the human capability to behave *within the society* by entering into relations in which the subject *can* or *cannot* be substituted. The subject has the power to refer data to herself, namely, to give an account of herself by means of personal data. This reference can be done either by the subject or by someone else (whether a human or an artificial agent), who is meant to act *on behalf of* the subject. This is the substitution made possible by the transformation of the trace into sign: the self is a reference that can become subject to someone else's identification. The self is identified and 'said' as if she were a third person. In this perspective, the grammar of the third person is much more than an instance of language: it is the possibility to make reference to someone else in the public domain. It is this regime of 'thirdness' or substitutability that allows someone to act or to speak *on behalf of* someone else.

In this sense, we must notice that the problem of privacy can no longer be confined to the risk of the *exportation* of personal data from the domain of the self towards the public domain. On the contrary, the problem of privacy concerns also the risk of *importation* of data within the domain of the self by who/what can act or speak on behalf of the subject. We refer to all cases in which the effects of someone else's power of reference can give an account of ourselves by means of either a concrete or an abstract substitution. In a Surveillance Society, this goes until the point that individuals can no longer defend a secret from the others, since, because of the importation of data, they can learn about themselves what they could even not know before. The ultimate form of substitution is, therefore, the dimension of abstraction, that is, the construction of abstract profiles (e.g., statistical), where everything is public in the sense of common to anybody. In this perspective, the problem is how to cognitively represent the extension of the category of *behalfness*, which defines the way in which personal data are no longer referred to someone by means of the process of self-identification or exportation, but through an automatic substitution or importation, which does not require a relation between me and the others. What difference if this imputation is made by machines instead of humans? This question brings us back to the issue of responsibility. In the autonomic computing system, autonomic managers seem to be entrusted with the crucial role of intermediaries between the two levels of decisions (selection and execution of goals) and, to this end, they are repeatedly associated with the term responsibility, which remains quite metaphorical. On the contrary, we judge responsibility is fundamental in a self-governing system for two reasons: it mediates between the levels of decisions and it is the dimension in which humans express their autonomy and subjectivity, since personal responsibility is what I cannot, by definition, be substituted for.

We should clarify this point philosophically. The regime of responsibility is not, in our view, deduced by some established general laws: of course, at a certain point, legal responsibilities are assigned according to existing norms. However, this point is never a starting point. Before being the content of laws, *responsibility* is a concrete human relation where one is called upon by someone to respond for something (not subject to anticipation). I draw this idea by Lévinas' philosophy of death. I become who I am in the moral relation to someone else's death: the other's death is part of my present but cannot be subject to anticipation, because of its unpredictability. This possibility interrogates and puts me into question (Lévinas 1997: 132–33): have I done everything to avoid her death? In this relation of responsibility, nobody can substitute me. Where I am personally called upon to respond to the other, there I can recognise myself as autonomous and free, since I may assume my responsibility for the other. It is the heteronomy of the call of the other that grounds my autonomy. In this perspective, autonomy is no longer based on the power of self-determination (the establishing of general laws) but in the moral tension of responsibility, which precedes freedom (Lévinas 1996: 83 and Rey 1997: 34–37): the 'I' is such, not because it is situated at the start of the chain of causation (as a legislator), but because it is *subject* to the relation of responsibility towards the other.

Here lies an essential point of a self-governing system: the concrete relation with the other is crucial, in order to develop my subjectivity. Otherness is what cannot entirely be anticipated, unless we destroy it. A system that is designed to fully anticipate its own environment risks to alter the possibility of entering into relation with the others. Our effort, as humans, to set general goals and directives, in order to design the autonomic system's power of anticipation, is likely to narrow the space of our relations with the others.[10]

The vital and significant relation with the others should not be made subject of an automated process, which seems to lessen the unpredictability and richness of our communications and relations, reduced to a system designed to fully anticipate its own environment (made also of personal data). This brings us back to the problem of privacy and identity, not only concerned with the protection of data but also with the construction of personal identity. We should reflect on how personal identity is constructed in an informational ambient, where personal data stem from multiple, distributed and thus competitive sources.

The 'competitive' self

Personal identity is a *polemological* concept. As Mathieu points out, 'privacy does not constitute an object but a relation, which cannot at its turn be objectified, of an object with an origin and an intention [. . .]. In other words, privacy is as well as law a protection of freedom, but as "freedom of" and not only as "freedom from"' (Mathieu 2004: 95). The violation of privacy does not entail the violation of an inner self, that would be deprived of her personal data or of the control thereof: a violation of privacy deprives the self of an essential capacity, that is, the ability of referring a content to herself (Mathieu 2004: 92–93), namely, of narrating her own history. This capacity is neither exclusive nor uncontroversial: the narration of the identity is irreducibly competitive.

Personal identity grows out of a contest: we have underlined this point as regards to the autonomic computing system, by noticing that identity is constituted of data that stem from multiple, distributed sources. This competitive dimension holds true remarkably when the notion of identity is evoked in conjunction with that of narrative. What is a tale? As has been noticed,[11] a tale is *the expression of a conflict that marks the passage of time*. This point can be illustrated through a mental experiment concerning autobiography.

Imagine two persons invited to write about the personal history of one of them: the first one is invited to write an autobiography, whereas the second is invited to write a biography. They are both confronted with the same material, that is, a set of personal data. Unless the improbable situation occurs in which they write the same tale, they are likely to produce two different stories. Therefore, the two accounts of the same life are competitive: they both aspire to be considered the best account of what is narrated. Furthermore, this competition is not necessarily biased in favour of autobiography, since in principle there is no reason to make it prevail over biography as the best account of the narrated self. We may discern two narrating selves: *the autobiographical self* stemming from the process of self-reference and *the competitive self* stemming from the process of hetero-reference. The identity of the narrated self grows out of the competition between the two tales and, hence, between the two selves that we might wish to reconcile. Since reconciliation may be difficult or out of reach, the identity of the narrated self is likely to emerge inasmuch as the competition is *fair*, that is, both the autobiographical and the competitive self are given at least equal opportunities in narrating a true history of the self.

Consider now the case that only one person is invited to write about her own personal history in an autobiography. The identity of the narrated self is constituted by a set of data, which consists in a selection of data. This selection is significant, because it gives, positively, an account of the self. Yet, the omission is significant, since it participates in giving, negatively, a true account of the self. According to a hermeneutical or psychoanalytical approach, the omission may be treated as being as significant as the selection is. Also in this case, we can discern an *autobiographical self*, whose tale is significant for what it narrates (the selected data), and an implicit *competitive self*, whose 'negative tale' is significant for what it excludes. Again the identity of the narrated self grows out of the competition between two instances of the same self. Here, the competitive self is part of the autobiographical self. Both in the first case, concerning the inter-subjective relation, and in the second case, concerning the structure of subjectivity, the identity of the narrated self is polemological, since it emerges out of the relation between *competing instances* of the self. However, this competition is not a blind dispute but implies the relation between these competing instances: the relation with the others (inter-subjectivity) and the relation with the other that inhabits each of us (subjectivity).

This means that the identity of the narrated self and hence privacy are assured inasmuch as the competition is fair, that is, the account of the self is given by competing instances according to fair conditions (*equal opportunities*). Competition between different instances (between multiple, distributed sources and traditional storage devices) cannot be totally excluded because it is inherent to the structure of the self. What matters is that the construction is based on fair, reciprocal conditions and communication between those instances. These conditions seem to be put at risk in the autonomic system: they seem confined to the initial division into levels of decision. Because of this sharp division devoid of a concrete intermediation it is difficult to say whether the treatment of personal data is driven according to conditions of fairness and to a process of communication: we are only said that the autonomic system has to be provided with 'appropriate measures' to protect privacy, without ascertaining who/what is responsible for their application.

This brings about two consequences. Firstly, fairness should be included among the philosophical premises of privacy, along with autonomy and dignity. The evolution of ICTs can alter or exclude the conditions of competition and communication, which should be evaluated in relation to their *standing of fairness*. The minimal level of fairness is assured if all instances are given at least equal opportunities in narrating the self, that is, when the competitive selves are not placed in the position to prevail over the autobiographical self. Secondly, fairness and communication are displaced when authors no longer have a grip upon the technological reality, which in turn has a tight hold upon them. The mutual implication, according to which 'if I "make" technologies; they, in turn, make me' (Ihde 2003: 20), should not be biased in favour of one of the terms of the relation. To the extent to which the relation of implication between the agent and the reality is mutual and fair, privacy can be assured, since its assertion does not only require that authors can protect their identity but also that they can *constantly* participate in the design of the technological reality that will co-constitute them (Pagallo, 2011).

The digital environment

In the present and final section we have to summarise what we have said so far. Our digital presence within the net is above all documental: when acting in the net, we leave traces of our actions. These traces are data that can be transformed into signs: the authors appear only after the traces left. The trace left and trailed becomes the point of reference to reconstruct an identity and a centre of legal imputation. This requires the transformation of traces into signs. This process of transformation and of reconstruction of personal identity is always competitive. Not only is this competition inherent in the structure of the self but also the conditions of such competition are likely to be transformed by the evolution of the enabling ICTs.

Many ICTs are enabling technologies meant to provide human existence with new possibilities. Not all these possibilities are endorsed by human practices. However, if endorsed, these possibilities are likely to be transformed into human powers: social, economic, legal, and political powers. This has a main consequence for society: the technological development driven by general decisions cannot establish by itself the goals to be pursued in a self-governing system. However, the technological development can modify the allocation of powers that exist in a society and hence it is able to alter the conditions of competition among different interests that aim to coexist in a self-governing system (Durante 2007b). Technology is unavoidably part of the conflict between countervailing instances or interests: we make technology that in turn makes us. The relation of mutual implication between subject and reality is assured as far as the conditions of competition and communication with the others are fair.

In the digital environment, identity and privacy are concerned with the main problem of self-identification, which is understood as the power of the self to refer data to herself, rather than as the assessment of the private or public content of data. This power to refer data to the self is competitive because of the status of human identity. The competition among different narratives is based on the materiality of the trace that allows us to have a hold on the technological reality that has a grip on the human reality. The mutual relation between the trace and the sign is a possible mediation between the agent and reality or, to refer to the autonomic system, between the two levels of decision. This relation is also necessary to reconstruct our personal identity growing out of the communication between multiple, distributed sources of data. In this perspective, we have to account for the questions raised at the start about the autonomic computing system as modelled on a self-governing system.

The general/specific levels of decisions

The trace inscribed is not by itself a sign, but can be transformed into a sign (relation of mutual implication). What does it mean as to the autonomic system? The autonomic system is organised in two levels as other self-governing systems: the level of general decisions made by 'legislators' and that of application of decisions made by 'executors'. In a self-governing system, general decisions (laws, goals or policies) are not by themselves plainly meaningful, capable of self-execution. This means that the general decisions established by humans being as legislators are only traces that are to be transformed into signs, in order to be applied. The generality of this level of decisions leaves to the executors the possibility to 'interpret' (and to alter) the general decisions. This is the reason why it is always necessary to have some control on the level of application: this control has to be, to some extent, conducted by humans, and it has to render the executors accountable and responsible. In a self-governing system, the transformation of traces into signs is a public, mediated, shared and controlled activity. The reference made in the vision of autonomic computing to a social *self-governing* system remains on this point quite controversial, since the clear-cut distinction between the two levels, in the lack of mediation between generality and specificity, is based on an old-fashioned construction of the subject as an absolute legislator, capable of determining the reality according to her plans. This vision of the subject is persuasive only in an epistemological representation of the relation between subject and reality as a perfect *correlation*, as if the autonomy of the agent consisted of a power of direct determination of reality. In contrast, the epistemological relation between subject and reality is one of *mutual implication*. Reality is only the trace of subjective determinations that need to be transformed into signs. The model of correlation, on which the autonomic systems impinges, is contestable: there are not two perfectly distinct poles, one fully determining and another one fully determined. What is 'determining' – in our language, what leaves the trace – is forever lost from the regime of the presence. It can only be referred to by transforming the trace into a sign, which indirectly relates us to what is meant to be 'determining': the sign substitutes what is lost and cannot be retrieved *as such*, if not in the limits of the substitution. What is concretely 'determining' (the pretended autonomy and agency of the subject) is the *process of transformation* of the trace into a sign, which requires mediation between the levels of decisions: it needs human interference – at least if one wants to speak of a *self-governing* system and not of a *self-executing system*, blind and devoid of accountability.

The need of mediation between levels of decisions

The transformation of the trace into a sign and the informational construction of the identity of the self is always competitive, that is, it demands the intervention of a plurality of agents and the communication between multiple, distributed sources of data. The trace can be taken as a sign but it does not cease to be a trace in its own materiality: it can be re-identified. Since the trace is inscribed, it determines the insertion of space within time and, hence, it gives people the possibility to make reference to this trace and to inhabit its space. The trace allows people to enter into mutual relation: this means that the relation and communication with the others is *always* necessary, in order to give full meaning to traces and to mediate between the two pretended levels of decisions of a self-governing system. If we renounce to the idea of an intermediation between decisions and their application, we should be ready to accept the idea of self-executing decisions, which leave no space to interpretation and unpredictability, to human intelligence and creativity. Relieving human beings from unwanted or unbearable tasks is a reasonable and wishful programme, but it is still different from conceiving general decisions as being self-executing. There is a difference that is not merely quantitative between designing artificial autonomous agents (positing problem as to their predictable behaviours) directed to specific goals and an autonomic system entrusted with general goals. In such a system, it remains quite unclear how autonomic managers, which play this crucial role of intermediaries, face the (legal) problem of responsibility for the unpredictable or unwanted outcomes that might arise from their operations.

The assignation of personal responsibility

The need of having a point of reference is fulfilled by the trace thanks to its inscription: the traceability and iteration of traces is the necessary condition to establish personal responsibilities, to reconstruct a coherent and reliable narration according to which one person may be called upon to respond to someone else. In a sense, in the autonomic system occurs the same, since the system could trigger a series of novel traces (inferred from the digital traces we leak) that we may read as signs to be imported or rejected, just like in our offline environments. In this perspective, the traceability of traces could assure the reconstruction of the operations deployed by the system. However, the assignation of responsibility for the operations that might engender violations of data remains quite metaphorical: it would be an imputation of effects to a source but not a dialogical process where someone is called upon to respond to someone else. In real life, this process is much more than the ascription of effects to a source: it is the dimension in which people, by being called upon to assume responsibilities towards the others, define the variable limits of their own autonomy and freedom. Saying that an autonomic manager is 'responsible for' means that they are 'charged with a task'. It does not refer to a process in which we are inter-subjectively confronted with the limits of our autonomy and freedom. If human agency is reduced to setting general goals designed to anticipate how the autonomic computing system is meant, at its turn, to anticipate its own environment, we are deprived of this dimension in which we are called upon to mediate between the general and the specific level of decisions and to assume personal responsibilities for the implementation of decisions.

From this it follows the need of designing the autonomic computing system in a way that does not unrealistically distinguish between a level where humans are exclusively concerned with the role of legislators and a level of execution from which humans are totally displaced: humans are not absolute subjects and full legislators, and machines are not mere executors. We have come to terms with this representation of human autonomy and agency. Humans' capability of anticipating themselves and their future is necessarily limited and irreducibly intertwined with the representation and the comprehension of reality they aim to govern: the ability of self-anticipation has to be constantly checked and corrected during the development of a self-governing system. General goals are similar to proposals (traces) addressed to a community of persons that is responsible for their transformation into binding directives (signs). This does not prevent people from trusting artificial autonomous agents by delegating them specific tasks. Trust does not necessarily require human touch. This differs, however, from delegating general goals to an autonomic system. In other words, I am not suspicious of artificial agents entrusted with delimited and specific goals but of the human attitude of thinking that general goals might be both the content and the limit of a self-governing system of decisions, which nobody can be called upon to respond personally for.

Notes

1 See www.research.ibm.com/autonomic/overview. All quotations concerning autonomic computing are drawn by the official IBM website. In particular see Kephart and Chess (2003).
2 On the notion of 'writing' as a mediation between law and technology see Hildebrandt (2008).
3 On this point, which in Lévinas is related to the constitution of subjectivity, you may see Durante (2006).
4 See also Floridi (2005: 194), with reference to what is traced: 'We are constantly leaving behind a trail of personal data, pretty much in the same sense in which we are losing a huge trail of dead cells'. See also Ferraris (2009).
5 Ferraris (2009: 34), Foucault (1979).
6 The public goes, according to Heidegger's *Das Man*, until what is impersonal and, eventually, anonymous. See on this Heidegger (2008).
7 On the notion of mediation you may see Durante (2009).
8 See www.research.ibm.com/autonomic/overview, 'The solution', where the notion of transparency is described with regard to the autonomic system: 'The system will perform

its task and adapt to a user's need *without dragging the user into the intricacies of its workings*'(we underline). This sounds to me as the opposite of transparency.
9 In the sense that Antoinette Rouvroy confers to this idea. See on this point Rouvroy (2011).
10 In a similar sense see, in the present book, Don Ihde's contribution (Ihde 2011).
11 I owe this idea to the writer Vincenzo Cerami (co-author of *Life is beautiful*).

References

Derrida, J. (1967) *L'Ecriture et la différence*, Paris: Le Seuil.
—— (1972) *Marges de la philosophie*, Paris: Les Editions de Minuit.
Durante, M. (2006) 'La 'subjectité' dans la phénoménologie de Lévinas', (104) *Revue Philosophique de Louvain*, 2: 261–87.
—— (2007a) 'The "Deepening of the Present" Throughout Representation as the Temporal Condition of a Creative Process', in A.-T. Tymieniecka, *Temporality in Life as Seen Through Literature*, Analecta Husserliana, The Yearbook of Phenomenological Research, vol. LXXXVI, Dordrecht: Springer: 285–310.
—— (2007b) *Il futuro del Web: etica, diritto, decentramento. Dalla sussidiarietà digitale all'economia dell'informazione in rete*, Torino: Giappichelli.
—— (2008) 'What Model of Trust for Networked Cooperation? Online Social Trust in the Production of Common Goods (Knowledge Sharing)', in T.W. Bynum, M. Calzarossa, I. De Lotto, S. Rogerson (eds), 'Living, Working and Learning Beyond Technology', *Proceedings of the Tenth International Conference Ethicomp*, Mantova: Tipografia Commerciale: 211–23.
—— (2009) 'Re-designing the Role of Law in the Information Society: Mediating between the Real and the Virtual', in Fernandez-Barrera, M., Gomes de Andrade, N., de Filippi, P., de Azevedo Cunha, M., Sartor, G., Casanovas, P. (eds), *Law and Technology. Looking into the Future*, Firenze: European Press Academic Publishing, 2009: 31–50.
—— (2010) 'What Model of Trust for Multiagent Systems? Whether or Not Etrust Applies to Autonomous Agents', *Knowledge, Technology & Policy*, 23: 3-4, 347-66.
Ferraris, M. (2007) 'Documentality or Why Nothing Social Exists Beyond the Text', in Ch. Kanzian and E. Runggaldier, *Cultures. Conflict – Analysis – Dialogue*, Proceedings of the 29th International Ludwig Wittgenstein – Symposium, New Series 3, Publications of the Austrian Ludwig Wittgenstein Society: 385–401.
—— (2009) *Documentalità. Perché è necessario lasciare tracce*, Laterza: Roma-Bari.
Floridi, L. (2003) 'On the Intrinsic Value of Information Objects and the Infosphere', (4) *Ethics and Information Technology*, 4: 287–304.
—— (2005) 'The Ontological Interpretation of Informational Privacy', (7) *Ethics and Information Technology*, 4: 185–200.
Foucault, M. (1979) 'Governmentality', in (6) *Ideology and Consciousness*, 5–21.
Ghanea-Hercock, R. (2007) 'Dynamic Trust Formation in Multi-Agent System', in *Tenth international workshop on trust in agent societies at the autonomous agents and multi-agent systems conference* (AAMAS 2007), Hawaii May 15, online at www.istc.cnr.it/T3/ trust.
Grodzinsky, F.S., Miller, K.W. and Wolf, M.J. (2009) 'Developing Artificial Agents Worthy of Trust: "Would you buy a used car from this artificial agent?"', *Proceedings of CEPE*, 26–28 June 2009, Greece.
Heidegger, M. (2008) *Being and Time* [1927] (trans. J. Macquarrie & E. Robinson), London: Harper.

Hildebrandt, M. (2008) 'Legal and Technological Normativity: More (and Less) than Twin Sisters', (12) *Techné*, 3: 169–83.

Ihde, D. (2003) 'Postphenomenology – Again?', *Working Paper n. 3*, The Centre for STS Studies, Aarhus.

—— (2011) 'Smart? Amsterdam Urinals and Autonomic Computing', in M. Hildebrandt and A. Rouvroy (eds), *Law, Human Agency and Autonomic Computing: The Philosophy of Law meets the Philosophy of Technology* , London: Routledge.

Kephart, J.O. and Chess, D.M. (2003) 'The vision of autonomic computing', IEEE Society, Manifesto.

Lévinas, E. (1996) *Humanisme de l'autre homme* [1972], Paris: Le livre de poche.

—— (1997) *Dieu, la mort et le temps* [1975], Paris: Le livre de poche.

Mathieu, V. (2004) *Privacy e dignità dell'uomo. Una teoria della persona*, Torino: Giappichelli.

Pagallo, U. (2011) "Designing Data Protection Safeguards Ethically", in D. Arnold and H. Tavani (eds), 'Trust and Privacy in Our Networked World', a special issue of *Information* (http://www.mdpi.com/journal/information/special_issues/ trust_privacy_networked_world).

Rey, J.-F. (1997) *Lévinas. Le Passeur de justice*, Paris: Michalon.

Rouvroy, A. (2011) 'Governmentality in an Age of Autonomic Computing: Technology, Virtuality and Utopia', in M. Hildebrandt and A. Rouvroy (eds), *Law, Human Agency and Autonomic Computing: The Philosophy of Law meets the Philosophy of Technology*, London: Routledge.

Serres, M. (2009) *Temps des crises*, Paris: Editions Le Pommier

Taddeo, M. (2008) 'Modelling Trust in Artificial Agents, a First Step Toward the Analysis of E-trust', in *Sixth European Conference of Computing and Philosophy*, University for Science and Technology, Montpellier, France, 16–18 June.

—— (2009) 'Defining Trust and E-trust: from Old Theories to New Problems', in (5) *International Journal of Technology and Human Interaction*, 2, April-June.

Chapter 6

Autonomic computing, genomic data and human agency

The case for embodiment

Hyo Yoon Kang

Slavoj Zizek remarked that '[w]ith biogenetics, the Nietzschean program of the emphatic and ecstatic assertion of the body is thus over. Far from serving as the ultimate reference, the body loses its mysterious impenetrable density and turns into something technologically manageable, something that we can generate and transform through intervening into its genetic formula – in short, something the "truth" of which is this abstract formula' (Zizek 2004: 25). Arguably, the privileged relationship between life and body has effectively been suspended since the ability to cultivate life *in silico* starting at the beginning of last century (Landecker 2007, Rheinberger 1997). But the loss of bodily density has been most acutely expressed in the so-called postgenomic era, which is marked by an intrinsic co-production of scientific artefacts and research questions with computational techniques and instruments within the biological metaphor of life as information or network (Haraway 1997, Kay 2000).[1] The notion of life as a 'pure information in computer networks, as robots, and as genetically engineered organisms' has been epitomised in computer simulations of digital artificial life as coded programs (Helmreich 2001: 125). The recognition of the dominant trope of information in postgenomics ought not be understood as falling into genetic reductionism, which equates genes to one-directional codes programming for biological life conceived as information processing machines. On the contrary, the availability of new technological tools has radically destabilised the concept of gene as a dominating epistemic framework of biological research in the twentieth century (Beurton et al. 2000, Moss 2003), making genetic data and information within the broader framework of life sciences and their social meaning even more important and valuable, precisely because their functions and importance are recognised to be much more complex than it has been assumed before the completion of the human genome sequencing in 2003.

From a legal theoretical perspective, the interesting point about postgenomic information to the issues of autonomic computing is that such data dissolve the distinction between the biological and the informational, and as a result radically destabilise the traditional boundary drawn between the human and the social: biology has gone computational, or rather it has merged into the practice of bioinformatics with the aim of producing and storing a mass of biological data and developing techniques for their analysis. In such an informational logic, human biological life itself becomes a legitimate object of computing, as Donna Haraway notes: 'human is itself an information structure' (Haraway 1997: 247). Emblematic of the Foucaultian notion of biopower, the biological and the social have literally become indistinguishable, both in their representational forms as digital data, as well as in their ontological character as material embodiments. Autonomous computing environments constitute a further progression within the process of increasing conflation of social and biological domains, as it results in the multiplication of forms of embodiment and locations of human subjectivity as the condition of human agency: the layers between the biological, virtual and real become increasingly permeable and enmeshed.

For legal operation, this development represents a considerable challenge for it marks the very dissolution of the traditional boundary it has drawn between the *tekhne* and *physis*, in which the mind represented the locus of form-giving agency that was applied on the matter of the body. Such a 'form/matter distinction . . .between rational consciousness and objective exteriority' was based on a post-Cartesian mechanistic understanding of nature, in which the body was regarded as part of nature that could be moulded and governed by human culture, of which law was part (Cheah 1996: 110, citing Collingwood). Long before the bioinformatics turn, feminist legal scholars have contested such a legal 'master discourse' (Shildrick 2005), which obfuscates the very social construction of the so-called 'natural' essence of human self. They identified and criticised the legal conception of human individuals as atomic, autonomous, rational, independent actors within controlled bodily boundaries by arguing that the nature/culture division obfuscates human bodily specificity and serves as a justification for social oppression by positing an essentialistic view of nature (Davies and Naffine 2001, Karpin 2005). Biotechnological advances have further crystallised and exposed the process in which concepts of human agency and personhood appear not as fixed, natural essences that precede their representations, but that it is literally the law as a social practice which maintains and stabilises the difference between subjects and objects, human and non-human, or nature and culture (Pottage 1998). Legal distinctions produce such *différances* in the sense that they are fully social meanings embedded in discursive and material practices (Derrida 1973). The property distinction between persons and things or the patent law distinction between invention and discovery are emblematic examples of such legal inventions, which have become strained in their internal logic and applicability in light of the ability to treat life as technology.

Far from being merely instrumental, legal boundaries in these contexts have been exposed as 'creative' techniques or 'conceptual fabrications' (Strathern 2004), by which law constitutes the very object to which it refers rather than representing an external reality. In Latourian terminology (Latour 1993), the inherently hybrid character of networks is 'purified' by legal categories such as personhood or invention, which simultaneously results in more nature and culture, or more of both humanness and artifice (Strathern 1996). In such a view, law itself represents a knowledge-making process at large (Riles 2000), and it is in this sense that law matters to the question of human agency in light of the challenges posed by the digital nature of socio-biological semantics and autonomous computing environments. For it is the legal conventions – fictional as they may be -- which stabilise and uphold the notion of human agency within the proliferating hybrid network of multiple agencies, the prospect of autonomic computing of bioinformatics data affords an interesting framework in which to reconsider the notion of the bounded biological self as the locus of human agency.

I understand the implications of the linkage between bioinformatics and autonomic computing on the legal conception of human agency to be compounded in the notion of embodiment – both conceived as a bodily, relational and situated experience, as well as a legal norm involving the question of representation. Embodiment in the first view stands for an experience, which is 'contextual, enmeshed within the specifics of place, time, physiology, and culture, which together compose enactment' (Hayles 1999: 196). The question for law is how to understand and treat information as legitimate legal embodiments. As Alain Pottage pointed out, 'the notion of embodiment not only serves the cognitive function of "reifying" intangible artifacts, it also articulates an essential normative distinction between science and technology, between disinterested knowledge and instrumental research' (Pottage 2006: 87–88).

The aim of this chapter is to explore the notion of embodiment for a legal reconceptualisation of human agency in light of autonomous computing environments. First, I analyse a patent application in the field of bioinformatics, in which implicit legal assumptions about the biological body and technological embodiments have come to the fore. The central question guiding my analysis is: what kind of assumption on the relationship between socio-biological information and materiality underlies the patent law practice in bioinformatics? How is the notion of embodiment employed in patent law and to what effect? Then I explore an alternative understanding of the legal human agent as an embodied social agent. In light of the increasing representation of living processes in informational terms, I draw on the notion of embodiment, especially the one elaborated by Katherine Hayles, in order to reconceptualise human agency as a relational and embodied capacity to action. I argue that the recourse to embodiment can provide a means by which information can be re-attached to specific corporeal and technological materialities.

Notions of embodiment in the practice of patenting of genomic information

Emphasising the complex interrelation between biological life, its representation as an information process and its codification in law, Stefan Helmreich wrote: '[t]hese days, mapping the socionatural terrains of "life" requires knowing who owns what' (Helmreich 2001: 137). Rather than being value-neutral, the meaning and materiality of digital information is thick with existing cultural notions of kinship, race and gender (Helmreich 2001). Immaterial socio-biological information underlies specific epistemic assumptions and material constellations, as for instance in intellectual property law. As a result, it becomes important to examine the very conceptual basis by which 'life' is constituted as a contingent category of ownership (Foucault 1970).

In light of the inseparability of information from its material embodiment in the genomic network, how has patent law dealt with the challenge posed by the multifaceted fluidity of genomic information? Patent law's subject matter and main technique of purification is the concept of invention, by which the proliferating nature-culture network, or vector, is cut and distinguished (Sherman 1994, Strathern 1996, Pottage 1998). The patentability criteria of novelty, inventive step and industrial application[2] presume and, by their application, reinstate a nature/culture dichotomy, by which inventions are seen as products of (human) industry, technology, scientific endeavours that transform otherwise (non-human) 'natural' objects by application of techne and labour. The legal definition of invention transforms biological data and material into legitimated objects of intellectual property right and which, together with technoscientific institutions and instruments, *creates* and *stabilises* the difference between nature and artifice.

Patent law's reaction to the convergence between biology and computing has been mixed. In the United States, bioinformatic patents are issued most commonly on gene chips and microarrays, with the caveat that the function of the software has to be specified in relation to the implementing hardware (USPTO 1996).[3] The first patent applications on a whole genome in a computer-readable format were filed by the company Human Genome Sciences (HGS) in the mid-1990s on the entire genomic sequences of three microorganisms (*Haemophilus Influenzae, Mycoplasma genitalium* and *Methanococcus jannaschii*). Bostanci and Calvert's study of these patent files (2008) revealed that the patent examiner understood the genome as a chemical compound rather than as a computable information, in line with existing patent law interpretation of genetic material as a chemical matter (Eisenberg 2002). The patent examiner asserted that 'computer hardware is an entirely different technology from molecular biology' and that 'searches for nucleic acid molecules would never uncover art related to computer systems' (Bostanci and Calvert 2008: 115). For postgenomic practices, the value of bio-information lies in its ability to be simulated, modelled, rewritten and recombined (Parry 2004, Pottage 2006). However, genomic information is also the result of a vast material infrastructure. As Bostanci and Calvert (2008) point out, 'the genome *is* a computer-related invention', in the sense that, without the computer technology, there would be no genome sequences and interpretation. The scientific practice and object of genomics is inseparable from computing and its technological assemblage (Cook-Deegan 1994, Morange 2005).

The hybrid nature of socio-biological information did not become recognised in the case of HGS patents, because it did not fit into the patent classification of legitimate forms of embodiment, which distinguishes between chemical, mechanical and electronic kinds. Although Eisenberg (2002) writes that, in patent law, '"tangible" now seems to mean "useful" rather than something material', the decisions on the HGS patent applications resulted in the reconstruction of digital biological information as chemical embodiments with material histories rather than legitimating their informational utility. In an unexpected fashion, the patent classification and the legal representation of genomic information as chemical entities effected a revisualisation of the tangible infrastructure and material embodiment of an invention, which was reflected in the rejection of a patent right in the whole genome in a computer-readable embodiment.

The legal requirement of specific embodiment in the context of computable bioinformation elucidates the underlying legal conceptions of materiality. Whereas some forms, especially mechanical processes and chemical matters, are viewed as legitimate inventions, digital information per se cannot constitute an invention, even if it arises from human ingenuity and intervention as in the context of bioinformatics practices. Disembodied information is literally not taken as being legally 'material'. Some forms of embodiment are recognised as patentable artefacts (for example, isolated genetic sequences in different carriers than in their 'original' environment), whereas others are not ('naturally' occurring forms, such as human body or some plant forms).[4] The legally legitimate form of embodiment can be chemical or semiotically inscribed in a software embedded in mechanical hardware. Although patent law's requirement of embodiment is based on the theoretical premise of a separation between materiality and information (and the corresponding dichotomies between invention and discovery, or culture as opposed to nature), its practical necessity to fix visible, fictional demarcations for the purpose of identification of property boundaries requires information's embedded materiality rather than lending recognition to its abstract code.

As a result, whereas, in the practice of bioinformatics, 'information lost its body' (Hayles 1999: 2), the legal requirement of embodiment reintroduces and reattaches bioinformation to the technological bodies of biosciences. What is embodied is the technological assemblage from which the information was derived rather than the human subject to which an invention could be attributed. Through such a legal re-materialisation of information, patent law acts as a disciplinary force, which is not necessarily concerned with the normalisation of human subjects (Foucault 1977), but rather with the disciplining of technological objects and processes according to property norms: what becomes re-embodied in the context of bioinformatic patents is the *material technological process* through which human biological life has become digitally represented. Although aspects of the human embodied self could be represented as associated attributes of inventions or as moral considerations (European Patent Convention, Art. 53 (a), (c)), they are rather seen as being enmeshed in a network of human and non-human *actants* (Latour 1988) and do not occupy central epistemic positions in the patent law discourse.

Relational agency and embodiment

Patent law's embodiment of bioinformation as mechanical or chemical matters resulting from technoscientific processes delimits human and non-human agential capacities. Precisely by destabilising and dislocating the notion of agency from persons to things themselves, patent law articulates relational subjectivities that cannot be located within human bodily boundaries, but arise as results of nego-tiating the contested material-semiotic space between the human agent, commercial and technoscientific practices, and legal technologies (Kang 2004). In line with Bukatman's observation that 'the point of origin of the subject is shifted to meaningful interiority and consciousness-driven stability to a complex and shifting techno-cultural configuration' (Braidotti 2002: 245, discussing Bukatman 1993), patent law's embodiment requirement reflects such a distributed configuration of disembodied human agency within a broader social and material technological constellation.

Parallel to such a recognition of distributed agency, there has been continuing unease about how to categorise inventions that bear human characteristics, which has conveyed the sense in which disembodied genomic information has been perceived as a threat to the traditional notion of liberal humanist subject. The fear of losing the privileged human ontological position is reflected in the proliferation of reports on the ethical aspects of intellectual property law,[5] as well as by the designation of intellectual property law as a novel area of human rights.[6] What such appeals to human rights or universal human dignity fail to take into account, however, is that the problem of disembodiment, that has given rise to such dis-quietudes, has already been prefigured in the legal characterisation of the liberal humanist subject. Hayles describes such an underlying premise of immaterial human subjectivity succinctly: 'Identified with the rational mind, the liberal subject *possessed* a body but was not usually represented as *being* a body. Only because the body is not identified with the self is it possible to claim for the liberal subject its notorious universality, a claim that depends on erasing markers of bodily difference, including sex, race, and ethnicity' (Hayles 1999: 4–5). Such a separation between information and embodiment is reflected in the traditional legal conception of the human subject that understands the essence of human self, and thus of agency, to be contained in an autonomous, rational mind.

Therefore, if the human subject ought to be re-embodied in light of the cultural and technological advances which have left it increasingly disembodied, this aim is not helped by a recourse to or propping up of the original humanistic conception of the human subject as a bodiless autonomous subject. Instead, emphasis on notions of the material body and the lived-in experience of embodiment can provide an alternative narrative of human agency. Differently from a naturalistic and essen-tialist understanding of the body as moldable matter formed by human culture, the notion of embodiment envisages agency as arising from a process of mutual causal relationship between bodily materiality and sociocultural forms. From such a perspective, it is not only the mind, but also the situated body that has agential capacities and which is specific, multiple, dynamic and dense (Butler 1993, Grosz 1994). Such bodies are not docile as in the early Foucaultian meaning of biopower (Foucault 1977) and imprintable by practices of knowledge like blank sheets of paper (as conveyed in the narratives of artificial life), but are intrinsically enmeshed in a web of power relations as participating and resisting agents with a capacity for recalcitrance (Latour 1988, Grosz 1994).

At this point, I ought to clarify by what I mean by embodiment which I draw from Hayles' concept of 'body/embodiment' (Hayles 1993: 153–54). In such a twin concept, the body signifies a normalised construct. Embodiment represents human contextual and temporal experience, which never coincides exactly with the body, but is enacted by human consciousness' interaction with social processes. Hayles' understanding of embodiment stipulates an oscillating unity between the norm and corporeal experience, which has the effect of folding the environment into the integral constitution of the body itself, rather than positing a separation between them.

In order to understand the social and physical actuality of embodiment and its constitutive role in the formation of human subjectivity, it is helpful to examine Hayles' argument more in detail. Her analysis of the present 'posthuman' condition is based on the argument that human cognition is formed in a process of *enactment* between embodied human selves and their (technological) environments so that mental processes cannot be fully understood without their embodiments (Maturana and Varela 1987). Enactment, a notion developed further in detail by Francisco Varela in the context of autopoietic theory, stresses the role of embodied agency in an organism's material development by arguing that perception and cognition are continuously shaped by changes in an organism's environment rather than representing triggers to external events by closed self-referential systems. In Hayles' discussion of Varela's later works, she points out that the notion of enactment radically overthrows the boundaries of the liberal subject, exposing that 'they are revealed to have been illusion all along' (Hayles 1999: 156). On the basis of contemporary models of cognition, Varela, together with Thompson and Rosch, argues that the mind, which is regarded to be the privileged source of human agency in law, should be understood 'not as a unified, homogenous unity, nor even as a collection of entities, but rather as *a disunified, heterogeneous, collection of processes*' (Varela et al. 1991: 100, cited in Hayles 1999: 157, original emphases). However, differently from disembodied visions of the mind as computable information as advanced by Hans Moravec or Marvin Minsky, Varela – in accord with his former teacher and co-author Humberto Maturana who insisted that the body and mind form a 'unity' – criticises such cognitive models as missing the important link which constitutes living cognition in the first place: namely the shaping of agency by sensory and physical perceptions of the environment (Varela 1992). Varela does not understand human agency as a product of abstract information processes, but posits embodiment, conceived as an evolving physical process indissolubly coupled to experiences of the outside world, as its main premise. Rather than representing a medium or carrier of computable mental processes, embodiment is the central condition and constituent of human cognition.

For law, the above figurations of agency cause a significant disjuncture between the locations of human material body and the experience of embodiment. The notion of 'human' agency itself becomes obsolete in such a picture, but rather is revealed to have been an essentially hybrid social-biological agency, all along. The task for law is, then, to depict and account for an embodied human subjectivity, which is also inherently social and technological because it is produced at the junction between lived-in, specific bodily experience, corporeal matter and the material constellation of technoscientific knowledge practices. How can law reflect this dispersion of human-social subjectivity without falling into the danger of reducing humans into 'things' or to accidental products of technological constellations? Moreover, how can such a hybrid subject position be delimited and visualised?

The legal scholar, Gunther Teubner, argues that the law can distinguish between different degrees of agency, which do not need to coincide with human agency:

> 'Actants' and 'hybrids' in the emerging ecological discourse in politics need not to be equipped with full-fledged legal subjectivity in order to open new political dynamics. Multiple legal distinctions – distinctions between different graduations of legal subjectivity, between mere interests, partial rights and full fledged rights, between limited and full capacity for action, between agency, representation, and trust, between individual, several, group, corporate and other forms of collective responsibility – have the potential to confer a carefully delimited legal status to political associations of ecological actants.
> . . .Legal capacity of action can be selectively attributed to different social contexts.
>
> (Teubner 2006: 517).

Granted that there are different creative legal techniques, which can fabricate hybrid agencies, the question remains by what criteria the law recognises and attributes agential capacity to humans and other social actors: how should the 'legal capacity of action' be defined? In the context of autonomous computing and high computing environments, it is precisely the hybrid constitution of agency, which makes its legal delineation intensely political. Arguably, the link between computed information and bodily embodiment in information-intensive environments is no longer one of representation, but one of mutation in which the computed information becomes part of the extended and distributed cognitive system, in which human cognition also partakes and by which it is shaped. Such a constellation of distributed embodiments and cognitions calls for both an extended account of the legal (post)human subject and an expansion of scope of ethics beyond humans (Hayles 2009), as the present depiction of a rational, bodiless human subjects as the main social agents is by design unable to account for hybrid processes of cognitive and embodied enactments.

The current legal predicament can be characterised as follows: precisely because body and 'nature' have been understood as fixed givens in the legal postulation of human personhood, too much emphasis has been put on the disempowerment of the human self by mainly characterising it as a mechanical process of information. However, rather than representing simple vehicles for the expression of socio-biological information, bodies have always been dynamic and multiple. In the context of the currently dominant trope of computable information both as representations of life and as programmable living processes, the concept of embodiment can serve as a useful criterion in order to differentiate the multiple layers of materiality that underlie and interpret such information. It would entail the recognition that embodiments are highly complex, dynamic and specific processes (Hayles 1993) and are therefore not always transparent (Butler 2005) like readable and computable data. The incorporation of the embodiment process into the narrative of legal human subject would act both as a normative idea, as well as representations of the social formation of human cognition. There are three ways in which the notion of embodiment can be employed in law.

First, the concept of embodiment serves as an interpretive framework through which computable information and its impact on human perception are understood as a continuous, co-constitutive relation rather than as separate, independent processes. The emphasis on embodiment allows law to interpret experiences of information as an innately corporeal and sensory phenomena, which – given the high specificity of individual organisms and social experiences – will reflect the individual corporeal and emotional responses to information. The singularity of human experience as an embodied one allows law to locate and redraw the boundaries of the human subject, not as a 'stable, autonomous, self-sufficient, and independent' one, as Karpin criticised (2005: 205), but as an emergent material self, embedded in relations to the world. Particularly, in the context of computable biological information, the concept of embodment enables law to re-attach information back to human bodily materiality. Bearing in mind that human beings share genomic information which can be computed and modelled, but whose individual bodily expressions and matters vary considerably, a legal concept of embodiment can also reflect the increasing convergence in postgenomic science between genotype (heredity) and phenotype (particular individual properties) (Mueller-Wille and Rheinberger 2009) through the folding-in of shared genomic information into individual experiences of embodiment.

Second, emphasising that information is always embodied, contextual and interpreted, a legal criterion of embodiment effects an opening-up of the very material composition of technology, which generates such information and their meanings. As 'the meaning of information is given by the processes that interpret it' (Fredkin 2007, cited in Hayles 2009: 66), the emphasis on information's embodiment discloses the multi-layered processes of interpretations that data undergo in order to become information in the first place. With regards to autonomous computing environments, the crucial point seems to be that human agents are not the exclusive interpreters of information, but layers of computable processes are actively giving meaning by interpreting information. This means that technological processes ought to be taken seriously as meaning-generating agents and held to account. The example of the patent application for human genomic information in computer-readable format, which I discussed in the previous section, illustrated how the patent law requirement of embodiment has led to the specification and representation of the technological assemblage as the proper boundaries of the invention rather than identifying the intangible genomic information as the legal subject matter. Although patent law was not operating with the philosophical concept of embodiment in mind, its practical necessity to visualise and delimit the object of intellectual property right by a tangible identification effected an understanding that digital genomic information is produced by technological practices, which, in turn, are embodied by specific, material set-ups. Thus, a legal requirement of embodiment can serve as a technique by which intangible information can be re-attached and localised to the specific technological body.

The other side of the coin of patent law's embodiment of non-human, technological agencies is that the practice of patenting human genetic entities has become so contested, precisely because patent law's conception of inventive artefacts does not recognise human processes of embodiment, which are also engaged in a co-production of subjectivities with the technological lifeworld. Instead of representing multiple levels of agencies and consciousness, such as the human embodied consciousness and the not always noticeable levels of the 'technological unconscious' (Thrift 2004), patent law trots along the traditional division of persons and things, which roughly overlaps with the distinction between human and non-human. As humans, by legal orthodoxy, cannot be property, patent law transfers human subjectivity into a non-material, disembodied realm of the 'sacred'[7] as expressed in the notion of human dignity,[8] which has the adverse effect of disconnecting the human from the singular experience of embodiment. In order to redress such a lack of representation of processes of human embodiment in the context of patent law, it would be necessary to incorporate representations of human embodiments into the patent law discourse itself. This would imply that the traditional legal dichotomy between persons and things would have to be discarded and replaced with a finer distinction between different levels of human and technological cognition and agency. In such an alternative discourse, the uniqueness of human embodied cognition would be asserted as being more capable of emergent consciousness than mechanical processes, instead of not being represented, at all, which is the present case, whilst the complexity of human and technological interrelations could also be taken into account by the requirement of material embodiment of technological processes.

Third, the concept of embodiment can serve as a normative criterion for the legal boundary-drawing between human and non-human agencies. If we envisage the human agent as an embodied being rather than as a rational, self-sufficient agent, such a view necessitates the interpretation of computable information according to their embodiment-enhancing or denying capacities. As a result, human autonomy would need to be reformulated as a relational and embodied capacity. The feminist legal scholar, Jennifer Nedelsky, has argued that 'autonomy is a capacity that exists only in the context of social relations that support it and only in conjunction with the internal sense of being autonomous' (Nedelsky 1989: 7). Important in such a conception of relational autonomy is the 'internal sense of being autonomous': in light of the central process of embodiment, in which individual sensory perceptions and consciousness are based on bodily experiences of social relations, embodied human subjectivity is the precondition of the existence of autonomy rather than being a logical consequences of the latter. This notion of emergent and situated autonomy closely resembles the understanding of autonomy in later autopoietic theory, in which Varela and Bourgine stress the emergent and creative capacity of living organisms to 'shape a world into significance' by its actions: 'Autonomy . . .refers to [the living's] basic and fundamental capacity to *be*, to assert their existence and bring forth a world that is significant and pertinent without being pre-digested in advance. Thus the autonomy of the living is understood here both in regards to its actions and to the way it shapes a world into significance' (Varela and Bourgine 1992: xi, cited in Hayles 1999: 22–23).

In light of such an understanding of autonomy as relational and embedded capacity, the recognition of distributed cognition and agency in the context of information-intensive environments presents an apposite occasion for a legal reconceptualisation of human subjectivity and the distinction of information in relation to autonomy. For the legal purpose of distinguishing between different effects of information on human material embodiments, information can then be assessed in light of its composition as a unity of 'pattern /randomness'. As Hayles explains, pattern is 'defined as the probability distribution of the coding elements composing the message' whereas randomness denotes a disruption of such repetitive patterns, which has the paradoxical ability to re-organise information 'at a higher level of complexity' (Hayles 1999: 25). Tying the relational and embodied definition of autonomy and the understanding of information as pattern/randomness together, this implies that the very structure of pattern/randomness in coded information can enhance or restrict the forming of relations between human embodiments and their environment because information shapes the experience of human embodied cognition, as well as being shaped by the material organism itself. I suggest that the definition of embodied, relational autonomy could be fruitfully used as a yardstick by which to assess the composition of information in terms of their effect on the emergent flow of human autonomy.

In more concrete terms, this implies that law must ensure that informational environments allow for the emergence of complex subjectivities by assessing a programme's potential for randomness, so that unpredictable mutations between the human self and its environment can occur. As studies of complex systems by Varela (1992) and Kauffman (1993) have demonstrated, it is randomness which is the precondition for patterns to emerge in the first place and which allows for higher levels of creative complexity and flexibility (Hayles 1999: 286). Derrida called such non-calculable randomness an 'excess' or 'freedom':

> But in the machine there is an excess in relation to the machine itself: at once the effect of a machination and something that eludes machinelike calculation. Between the machinelike and the non-machine, then, there is a complex relation at work that is not a simple opposition. We can call it freedom, but only beginning at the moment when there is something incalculable. . . .The event

– which in essence should remain unforeseeable and not programmable – would be that which exceeds the machine.

(Derrida and Roudinesco 2004: 50)

In order to construct human autonomy amidst a web of computed information, the law needs to distinguish between the kind of programmes in which no sufficient randomness of embodied relations is granted and therefore threaten the human capacity for 'random' association and the kinds of computed information which allows for the emergence of multiple, open readings and incorporations of meanings into specific contexts. The legal construction of human autonomy in relation to autonomous computing environments, especially in relation to the use of computable bio-social information, ought to codify randomness as a *value* and *precondition* for the formation of autonomous subjectivity. The formation of patterns may follow as a consequence of the quantity and quality of randomness, but it is the safeguarding of randomness, which enables the very human experience of embodiment as a condition of being in the world.

Conclusion

Amidst the apparent disappearance of bodily density and opaqueness brought about by the representation and intervention of living processes as computable information, the recourse to the technique and value of embodiment can help to reconceptualise human legal subjects as material bodies with embodied minds rather than understanding them as computable minds with absent bodies. The concept of embodiment re-establishes and addresses the singular and shifting character of corporeality of human agents as recalcitrant and relational, and thus provides an important alternative to an understanding of human agency as exertion of one's will over a docile body. In the latter view, such as in the context of property, the question of agency has been traditionally framed as a question about 'who owns what' along the stipulated division between humans and things. However, if the law envisaged the legal human agent as an embodied one, who is fully embedded within a web of multiple social and non-human technological agencies, it would be able to reformulate that question to: 'who/what owns what/who?'. Such a revised understanding of property relationship would sketch a more co-constitutive picture of interaction between humans and non-humans. The concept of embodiment, therefore, provides an alternative narrative to the abstract idea of living processes as disembodied information. Also with regards to non-human entities, the requirement of embodiment allows for the examination and representation of the material technical constellation from which information arises. Thereby the law can ply open the embodied processes by which an information becomes interpreted and is given meaning – an aspect which is often forgotten, as if information was without an origin and context. The case of a patent application for an invention containing computable genomic information presented such an example of how the legal requirement of embodiment effected a re-materialisation of information.

I have argued that the concept of embodiment serves as a useful normative criterion, by which computable information can be assessed in their autonomy-enhancing or –denying capacities. In order to do so, however, the law needs to reconceptualise and redistinguish human autonomy and agency within the web of technological and social relations rather than understanding human agents as disconnected, autonomous beings, independent of social relations, who exercise their will with unfettered agency. In the context of the complex epistemic and normative questions posed by autonomous computing, it is not only the prospect of the technology that unsettles human agency, but its understanding in an essentialist and isolationist way that prevents a more complex and richer distinction of human capacity for autonomous actions within the technoscientific environment.

Embodiment is what makes us human. It is not only our rational capacity, but the context and specificity of embodiment that distinguish us from things: 'human cognition remains distinct from thing-hood because it arises from embodied contexts that have a biological specificity capable of generating consciousness as an emergent phenomenon, something no mechanical system can do' (Hayles 2009: 66). The concept of embodiment introduces randomness and incalculability into the legal narrative of human agent. Thereby it can provide the basis on which relational and situated autonomy can develop.

Notes

1   In operational terms, genomics denotes the study of the molecular structure of whole genomes by investigating their DNA structure. Postgenomics represents the phase after the successful sequencing of whole genomes in the 1990s, which has been increasingly marked by questions about genomic expressions and functions to which there have been no determinate answers (Mueller-Wille and Rheinberger 2009: 9).

2   European Patent Convention, Art. 57.

3   In Europe, the jurisprudence relating to bioinformatics patents is contested. *Kirin-Amgen v. Transkaryotic Therapies* confirmed the patentability of methods using genetic information to be embodied in different carriers ([2002] RPC 187). Bioinformatics patents conceived as computer-implemented inventions could, in principle, be patentable as long as they make a technical contribution (*Viacom/Computer-related Invention*, T208/84 [1987] *EPOR* 74). However, the area is mired in controversy as European Parliament's rejection of the proposed EU Software Patent Directive in 2005 has shown (Draft Directive (COM/2002/0092)).

4   *Howard Florey/Relaxin* [1995] EPOR 541.

5   A number of reports on the bioethics of patenting inventions of human origin have been published. See for example, President's Council on Bioethics (US) www.bioethics.gov/background/workpaper8.html; German bioethics committee's opinion: www.ethikrat.org; French National Bioethics Advisory Commission's opinion°064: www.ccne-ethique.fr/?langue=2; the UK Human Genetics Commission's page on: www.hgc.gov.uk/Client/Content.asp?ContentId=362.

6   The UN High Commissioner of Human Rights' taskforce on bioethics http://www2.ohchr.org/english/issues/bioethics/. All links visited on 13 December 2008.

7   See Agamben (1998) for the notion of the sacred as falling outside the legal realm. On the quasi-religious aura of genetic language, see Kay (2000).

8   See Pottage (2002) for an exposition of the employment of the concept of human dignity as a legal blackbox.

References

Agamben, G. (1998) *Homo Sacer: Sovereign Power and Bare Life,* Stanford, CA: Stanford University Press.

Beurton, P. J., Falk, R. et al. (eds) (2000) *The Concept of the Gene in Development and Evolution*, Cambridge: Cambridge University Press.

Bostanci, A. and Calvert, J. (2008) 'Invisible Genomes: the Genomics Revolution and Patenting Practice', (39) *Studies in History and Philosophy of Biological and Biomedical Sciences*: 109–19.

Braidotti, R. (2002) *Metamorphoses: Towards a Materialist Theory of Becoming*, Cambridge: Polity.

Bukatman, Scott (1993) *Terminal Identity. The Virtual Subject in Post-modern Science Fiction*, Durham, NC: Duke University Press.

Butler, J. (1993) *Bodies that Matter: On the Discursive Limits of 'Sex'*, New York: Routledge.

—— (2005) *Giving an Account of Oneself*, New York: Fordham University Press.

Cheah, P. (1996) 'Review: Mattering', (26) *Diacritics* 1: 108–39.

Cook-Deegan, R. (1994) *The Gene Wars*, New York & London: Norton & Co.

Davies, M. and Naffine, N. (2001) *Are Persons Property?* Aldershot: Ashgate Dartmouth.

Derrida, J. (1973) *Speech and Phenomena*, Evanston, IL: Northwestern University Press.

Derrida, J. and Roudinesco, E. (2004) *For What Tomorrow. . .A Dialogue*, Stanford, CA: Stanford University Press.

Eisenberg, R. (2002) 'Molecules vs. Information: Should Patents Protect Both?', (8) *Boston University Journal of Science and Technology Law* 1: 190–217.

Foucault, Michel (1970) *The Order of Things*, New York: Random House.

—— (1977) *Discipline and Punish*, London: Penguin.

Grosz, E. (1994) *Volatile Bodies: Toward a Corporeal Feminism*, Bloomington, IN: Indiana University Press.

Haraway, D. (1997) *Modest_Witness@second_Millennium. Female Man_Meets_ Oncomouse*, London, New York: Routledge.

Hayles, N. K. (1993) 'The Materiality of Informatics', (1) *Configurations* 1: 147–70.

—— (1999) *How We Became Posthuman*, Chicago, IL: University of Chicago Press.

—— (2009) 'RFID: Human Agency and Meaning in Information-Intensive Environments', (26) *Theory, Culture, Society*, 2–3: 47–72.

Helmreich, S. (2001) 'Kinship in Hypertext: Transsubstantiating Fatherhood and Information Flow in Artificial Life', in S. Franklin & S. McKinnon (eds) *Relative Values,* Durham & London: Duke University Press: 116–43.

Kang, H. Y. (2004) 'Identifying John Moore: Narratives of Persona in Patent Law Relating to Inventions of Human Origin', in Glasner, P., Atkinson, P., & Greenslade, H. (eds) *New Genetics, New Social Formations*, London: Routledge.

Karpin, I. (2005) 'Genetics and the Legal Conception of Self', in M. Shildrick and R. Myktiuk (eds) *Ethics of the Body*, Cambridge, MA: MIT Press.

Kauffman, S. (1993) *The Origins of Order*, Oxford: Oxford University Press.

Kay, L. (2000) *Who Wrote the Book of Life?* Stanford, CA: Stanford University Press.

Landecker, H. (2007) *Culturing Life. How Cells became Technologies*, Cambridge, MA: Harvard University Press.

Latour, B. (1988) *Science in Action*, Cambridge, MA: Harvard University Press.

—— (1993) *We Have Never Been Modern*, Cambridge, MA: Harvard University Press.

Maturana, H. and Varela, F. (1987) *The Tree of Knowledge*, Boston, MA: New Science Library.

Morange, M. (2005) *Les secrets du vivant: Contre la pensée unique en biologie*, Paris: La Découverte.

Moss, L. (2003) *What Genes Can't Do*, Cambridge, MA: The MIT Press.

Mueller-Wille, S. and Rheinberger, H.-J. (2009) *Das Gen im Zeitalter der Postgenomik*, Frankfurt: Suhrkamp.

Nedelsky, J. (1989) 'Reconceiving Autonomy: Sources, Thoughts and Possibilities', (1) *Yale Journal of Law and Feminism*, 1: 7–36.

Parry, B. (2004) *Trading the Genome*, New York: Columbia University Press.

Pottage, A. (1998) 'The Inscription of Life in Law: Genes, Patents and Biopolitics', (61) *Modern Law Review*, 5: 740–65.

—— (2002) 'Unitas Personae: On Legal and Biological Self-Narration', (14) *Law and Literature*, 1: 275–307.

—— (2006) 'Materialities in Law and Life', (15) *Paragrana*, 1: 81–99.

Rheinberger, H.-J. (1997) *Towards a History of Epistemic Thing*, Stanford, CA: Stanford University Press.

Riles, A. (2000) *The Network Inside Out*, Ann Arbour, MI: University of Michigan Press.

Sherman, B. (1994) 'Governing Science: Patents and Public Sector Research', (7) *Science in Context*, 3: 515–38.

Shildrick, M. (2005) 'Beyond the Body of Bioethics', in M. Shildrick and R. Myktiuk (eds) *Ethics of the Body*, Cambridge, MA: MIT Press.

Strathern, M. (1996) 'Cutting The Network', (2/3) *Journal of the Royal Anthropological Institute*: 517–35.

—— (2004) 'Losing (Out on) Intellectual Resources', in A. Pottage and M. Mundy (eds) *Law, Anthropology, and the Constitution of the Social*, Cambridge: Cambridge University Press: 201–33.

Teubner, G. (2006) 'Rights of Non-humans? Electronic Agents and Animals as New Actors in Politics and Law', (33) *Journal of Law & Society*, 4: 497–521.

Thrift, N. (2004) 'Remembering the Technological Unconscious by Foregrounding the Knowledges of Position', (22) *Environment and Planning D: Society and Space*, 1: 175–90.

United States Patent and Trademarks Office (1996) 'Examination Guidelines for Computer-Related Inventions', *Federal Register*, 61: 7478–92.

Varela, F. (1992) 'Making it Concrete: Before, During, and After Breakdowns', in J. Ogilvy (ed.) *Revisioning Philosophy*, Albany, NY: State University of New York Press.

Varela, F. and Bourgine, P. (eds) (1992) *Toward a Practice of Autonomous System*, Cambridge, MA: MIT Press.

Varela, F., Thompson, E., Rosch, E. (1991) *The Embodied Mind*, Cambridge, MA: MIT Press.

Zizek, S. (2004) *Organs without Bodies*, New York, London: Routledge.

Chapter 7

Technology, virtuality and utopia

Governmentality in an age of autonomic computing

Antoinette Rouvroy

Introduction

> The machines, structures, and systems of modern material culture can be accurately judged not only for their contributions to efficiency and productivity and their positive and negative environmental side effects, but also for the ways in which they can embody specific forms of power and authority.
>
> (Winner 1986)

This chapter attempts to identify the repercussions for the constitution, experience and understanding of human personality and legal subjectivity, of the increasingly statistical governance of the 'real' ensuing from a convergence of contemporary technological and socio-political evolutions. Epitomised by the rise of autonomic computing in the sectors of security and marketing, this epistemic change in our relation to the 'real' institutes a specific regime of intelligibility of the physical world and its inhabitants. This new 'perceptual' regime, it will be argued, affects a specific and essential attribute of the human subject, which may be called his 'virtuality' (as opposed to 'actuality'). This 'virtuality', which acts as preserve for individuation over time, presupposes the recognition of 'differance' (being over time) and potentiality (spontaneity) as essential qualities of the human being. This virtual quality of the self, being a precondition to the experience of 'utopia' (spaces without location, according to Foucault), also conditions cultural, social and political vitality. Reflecting on the potential impacts of autonomic computing on human personality and legal subjectivity in terms of the governmental rationality these new technological artefacts actualise allows for a normative evaluation of the impact of autonomic computing on both individual self-determination and collective self-government.

The subject itself is highly uncertain. 'Autonomic computing' *per se*, is difficult to circumscribe as an *object* for legal theoretical inquiry. IBM, which first coined the term, explicitly acknowledges that 'the definition of autonomic computing will likely transform as contributing technologies mature'. IBM nevertheless lists eight defining characteristics for it, and presents the vision of 'computer systems that regulate themselves much in the same way our autonomic nervous system regulates and protects our bodies'. The defining capabilities of autonomic computing include self-knowledge (the system must somehow know itself and be able to identify its own components), autonomic and dynamic self-reconfiguration and adjustment, constant optimisation of its own working, self-prevention and reparation of malfunctioning caused by internal or external events, detection of and protection from attacks against the system's security and integrity, context awareness and autonomic adaptation of itself or even the environment to the circumstances, ability to anticipate and optimise resources consumption while keeping its complexity hidden. It must marshal I/T resources to shrink the gap between the business or personal goals of the user, and the I/T implementation necessary to achieve those goals – without involving the user in that implementation.[1]

As of today, autonomic computing is nothing more than a 'vision', which is not (or not yet) embodied in any specific 'artefact', scenario or application that would give rise to *actual practices* from which to start our study. As a 'vision', or a 'paradigm shift',[2] autonomic computing is aimed at facilitating and enhancing the functioning of a wide variety of information systems, going from the traditional laptop to the most complex computer-sensors networks one may imagine being involved in futuristic scenarios of ambient intelligence. The prospective stance one is unavoidably caught in does not allow for any clear view of the future.

Other difficulties arise from the radical instability of the concepts of human identity (or human personhood) and legal subjectivity. Human identity appears a concept continuously expanding in scope (and whose expansion has been received as the hallmark of civilisational progress) but remains definitionally uncertain. Legal subjectivity is not a firmer concept, obviously: both assumed and constituted by law,[3] it appears irremediably self-referential, or enclosed in a positivity that can never completely be relied upon.

Assessing how our notions of gestures and agency, subjectivity and identity mutate in the presence of real time, dynamically varying media managed by autonomic computing,[4] and how legal and moral responsibility must be understood in circumstances where self-awareness and intentionality appear somewhat dispersed, dissolved or distributed in a human-technological network, is obviously crucial. Yet, it is not my intention to assess whether these new information systems exhibit the requisite properties of agents and, if they do, whether granting them the status of agents would curtail agency, identity, autonomy in humans,[5] nor to inquire about the more recently typologised types of 'cyborg intentionality' 'which all involve specific blends of the human and the technological' (Verbeek 2008: 387–95). In this chapter, the mere ambition I have is to slightly displace the point of view from which to consider the issues that have so far been addressed in order to inscribe the debate in the epistemic and assumingly political context of the day.

My departure point is not a contemplation of how human *subjects* actively interact with autonomic computing systems (there are currently no actual instances where this happens except in laboratory, experimental conditions where prototypes are being developed), but how human subjects are taken as *objects* of observation, classification and forward-looking evaluation (Gandy 2010) by such autonomic systems, and what the consequences are of the 'production' of such statistically based knowledge. Said in other words, I wish to reflect on how these autonomic machines *translate* or *transcribe* the physical world, its inhabitants, their trajectories, behaviours, actions, choices, preferences, attitudes. . .Because there is no neutral transcript of the 'real', I wish to identify the underlying 'bias' governing the regime of visibility and of intelligibility implemented in this way.

Autonomic computing and governmentality

The questions I am concerned with are the following: what are the specificities of the new modes of intelligibility of the 'real', or of the new rationality that such technologies inaugurate? What 'axial principles'[6] does autonomic computing serve? The politically relevant question is thus: what is the kind of power that the new regimes of (in)visibility and intelligibility accompanying the deployment of such technologies are aimed at and/or are capable of bringing forward? To what type of governmental rationality are these regimes instrumental? And, finally, what impacts would the deployment of such artefacts have on the processes of subjectivation and socialisation, and on the collective capacity to invent new political and social ways of life? I realise that discussing all this in a single chapter can only be justified because a book length meditation would be nearly as inadequate.

The following reflections are much influenced by Foucauldian scholarship and 'governmentality studies', as an almost instinctive point of entry into what is to me a new territory of inquiry consists of considering the impact of autonomic computing (embedded as it is aimed to be in systems of ubiquitous computing and ambient intelligence)[7] not *directly* on our understanding of human identity and legal subjectivity, but to see this impact through the transformations of knowledge and power that such technological developments implement. The perspectives suggested by the notion of autonomic computing *per se*, 'an approach to self-managed computing systems with a minimum of human interference', inspired by the human 'body's autonomic nervous system, which controls key functions without conscious awareness or involvement' raise fascinating and troubling issues, but, I wish to argue, these issues do not have much to do with the quasi-organic model of development and maintenance envisioned in autonomic computing, which humans have always shared with all other living organisms without having felt threatened in their specificity as human beings. The issues I am concerned with rather relate to the 'regimes of truth' (Foucault 1980: 93), the categorisations and (sometimes performative) predictions these systems are capable to establish, maintain and propagate through a series of applications ranging from security to entertainment, passing by marketing, health management, etc. Because these 'truth regimes' will result from technological (rather than human) observation, detection, classification and forward-looking (and thus predictive rather than purely descriptive) evaluation processes, individuals, diversely apprehended through the prism of profiles built on numbers and data, will not retain much power over their recognition, interpellation, classification by and within the systems.

> Our capacity to reflect upon ourselves, to tell the truth about ourselves, is correspondingly limited by what the discourse, the regime, cannot allow into speakability.
>
> (Butler 2005: 121)

Fragmented as they will be into a myriad of 'correlatable' data and aggregated with others with whom they do not share anything more than the simple fact of having exhibited similarly correlated biographical, behavioural, or other elements, the profiled individual will not necessarily be able to contest or resist the autonomic assignation of profiles and the practical consequences ensuing in terms of access to places, opportunities, and benefits. This attests, in a radical manner, of the fact that identity, that which results from an identification process, can only be accounted for from a perspective which is not that of the subject himself, but of others.[8]

> Through the lens of representations thrown off by these practices, individuals, once understood as moral or rational actors, are increasingly understood as locations in actuarial tables of variations. This shift from moral agent to actuarial subject marks a change in the way power is exercised on individuals by the state and other large organizations (. . .) The effects can be discerned on the way we understand ourselves, our communities, and our capacity for moral judgment and political action.
>
> (Simon 1988: 772)

What I am concerned with is precisely the relation between, on the one hand, the process through which the physical world and its inhabitants are made visible and meaningful, through which states of affairs are seen and evaluated, through which evidences are produced and given, in a 'world of autonomic computing', and, on the other hand, practices of 'government', where

> '[G]overnment' [does] not refer only to political structures or to the management of states; rather, it [designates] the way in which the conduct of individuals or groups might be directed. (. . .) It [does] not only cover the legitimately constituted forms of political or economic subjection but also modes of action, more or less considered or calculated, which [are] designed to act upon the possibilities of action of other people. To govern in this sense is to structure the possible field of action of others.
>
> (Foucault 1982: 790 and Foucault 1994: 635–65)

I thus refer to governmentality in a Foucauldian (Foucault 1991a) sense as the 'conduct of conduct', identifying how the new regimes of visibility and intelligibility implemented by the considered technologies impact on how we conduct ourselves, how we attempt to conduct others, and how others attempt to control our conduct.

Governmental rationality [is] a way or system of thinking about the nature of the practice of government (who can govern; what governing is; what or who is governed), capable of making some form of that activity thinkable and practicable both to its practitioners and to those upon whom it was practiced.

(Gordon 1991: 3)

At a time where the next step in the development of our information society may be a turn towards autonomic computing, I believe that borrowing from the governmentalist perspective[9] may be highly suggestive and helpful to assess why, and, above all, *at what price* autonomic detection, classification and forward-looking evaluation would gradually assist or even replace human observation.

Is the turn towards autonomic computing a 'natural' or an 'ideological' gesture?

Yet, neither ubiquitous computing, ambient intelligence or autonomic computing *per se* do seem to have awoken the interest of scholars involved in governmentality studies, despite the obvious impact that such information infrastructure may have on how power is being exercised on individuals and populations, through the new regimes of visibility and intelligibility of the repartition of risks, merits, abilities, deserts, opportunities, propensities, etc. they instaurate. One reason for this might be that autonomic computing, and the applications it is aimed to sustain, appear somewhat 'natural', with the connotations that such 'naturality' entails in terms of political and epistemological neutrality.

Presented as a technological solution to the growing complexity of the information technology infrastructure, the concept of autonomic computing coined by IBM seems to indicate the '*natural* next step' (although advertised as a 'paradigm shift') in the development of computer science. By virtue of having its inspiration in biological ontogeny[10] ('a systemic view of computing modelled after a self-regulating biological system'), where computers would take care of themselves and of their own development, the concept of 'autonomic computing' suggests a reduction of the distance separating the domain of 'artefacts' from that of 'nature'.

The word *nature* has always been ambiguous though. The Greek notion of *physis*, for example, is broader than the French notion of *nature*, whose Latin origin links to the verb *naître* in French (nascor) and to a notion of *natality* or genesis of things. By contrast, the Aristotelian notion of nature or *physis* involves – rather than a fixed origin – an innate thrust towards alteration, transformation, metamorphoses, a movement guided by an internal teleology (Aristotle 1979: 366). The Continental tradition seems attached to a concept of nature definitionally opposed to the teleology of artificiality; a concept of nature as something spontaneous rather than created, and that attaches value to the integrity of the spontaneous or inherent teleology of organisms, given in the past. Typically, the French romantic conception of nature tends to consider that what is natural is what has been generated without artificial, human intervention. All this cannot give any account of the proliferating 'hybridity' (Latour 2005).

Anyway, the gradual withdrawal of human intervention and the correlative increase of computers' autonomic capacities would allow them to become the autonomic (un)conscious 'brain' of a variety of increasingly prosthetic (functioning as 'prosthesis' for human beings) information systems. Such integration in systems of 'smart environments' with which the 'user' interacts 'naturally' and invisibly, added to its growing self-sufficiency, may reinforce the impression that the turn towards autonomic computing, and the increasing reliance one places in such systems, are quasi-natural evolutions of computer technologies and, arguably, also of our own species. After such endorsement, the remaining questions would only be whether, to what extent and with what consequences these new autonomic artefacts are taking over some of the attributes that were previously thought of as distinctive of human identity and legal subjectivity, whatever one thinks these attributes are (agency, intentionality, free will, emotions, etc.).

Yet, unlike living organisms, technologies never result from a spontaneous germination, but follow a teleology of artificiality. Even as machines become increasingly autonomic and 'intelligent', they remain dependent – be it only for their existence – on an initial design, intention, conception, script or scenario, and are from the start (whatever the shape they may actually take afterwards) embedded with their designers' conscious or unconscious visions of the world, and projections or expectations of what the future will be or should be, how human beings 'normally' or 'expectedly' behave, etc.:

> Designers define actors with specific tastes, competences, motives, aspirations, political prejudices, and the rest, and they assume that morality, technology, science and economy will evolve in particular ways. A large part of the work of innovators is that of *inscribing* this vision of (or prediction about) the world in the technical content of the new object. I will call the end product of this work a 'script' or a 'scenario'.
>
> (Akrich 1992: 208)

Technologies are always designed with a specific purpose in mind, in the context of specific problems and applications. One cannot blind oneself to the recent transformations of the modes of knowledge production, increasingly oriented by agenda and interests of funding agencies (Stengers 2002). These agencies identify *what the problems are* for which technological solutions must be found. Michel Foucault emphasised how a given solution to a given problem is only ever constructed according to how the problem is perceived in the first place, though a 'work of thought', a process of 'problematisation'. This 'problem-setting' (Finlayson 2006: 541–57) activity or this 'problematisation' has political implications. It is of course a truism to say that the development of technology is never random nor in any sense natural, but responds to the specific needs that manage to federate enough political and economic support to appear worth developing.[11]

Not endorsing what is presented to us as the next paradigm shift in computer science as something 'natural' (in the popular sense of the term), nor the correlated connotations of political and ideological neutrality this reference to nature still (misleadingly) entails, paves the way for understanding the problematisation[12] – 'a process by which a putative problem is seen as requiring special attention especially by government' (Welch 2008: 229) – that brought the notion of autonomic computing to the fore (Foucault 1991b: 381–90).

Moving towards autonomic computing is not a self-supporting technological paradigm change, bringing a purely technological solution to a purely technological problem and causing, as collateral or side effects, fascinating uncertainties with regard to the meaning of human identity and legal subjectivity. To what 'problems' is autonomic computing intended to bring a solution? How are these problems selected and identified? Why are these problems sufficiently high on the list of priorities as to make autonomic computing appear the 'natural next step' to go? The question – *why?* – immediately refers to problematisation.

Towards a statistical (or actuarial) governance of the 'real'

> What is real – if something like that can ever be supposed to exist in itself – does not matter; what matters is what is taken as real and in modernity what is taken as real is statistically recorded.
>
> (Skouteris 2004: 15)

Ubiquitous and autonomous computing, multimodal observation, ambient intelligence and all these new technological *infrastructures* purporting to make our life safer, easier, more efficient and enjoyable are the next step in the colonisation of the physical world by digital technology. They enrich our daily life's cognitive experience with dynamic and individualised informational content. Their celebrated capacity to detect, sort, evaluate and, most importantly, predict our desires and preferences, needs and propensities, and to customise and adjust deliveries, services and offers to our individual profile as if it knew us better than ourselves spares us time and discomfort. Their aptitude to target more accurately and objectively the individuals whose trajectories and attitudes put at a higher than average probability of committing a criminal offence or being involved in some way or another in a terrorist attack, allowing more selective security screenings and leaving the 'good guys' in peace, renders counter-terrorism policies less obtrusive to citizens' everyday life. The learning system ends knowing what your needs are, and who the bad guys are. Unobtrusively, it renders your environment responsive to your unique personality (yes, you are unique, and the system will reassure you on that point) whilst eliminating most frictions with the unexpected, unpleasant, time-consuming, tiresome aspects of choices or routine security checks.

In order to perform their tasks as intelligent interfaces or smart mediators (and, possibly, agents) between human users and the humanly untameable complexity of the global digital and physical universe, and to deliver their individualised, dynamic functionalities (whatever these are), the new information infrastructures 'translate' or 'transcribe' the physical space and its inhabitants (that's us) into constantly evolving sets of data points. The optimal functioning of this mode of statistical intelligibility presupposes the non-selective collection of as-much data as possible, *a priori* independent of any specific finality. At odds with the modern ambitions of deductive rationality linking observed phenomena (that is, phenomena previously selected, on explicit or implicit criteria of *interest*, as objects for obser-vation and analysis) to their causes, the rise of autonomic computing attests to a broader epistemic shift, the new 'perceptual regime' appears to follow an inductive (rather than deductive) logic. Indifferent to the *causes* of phenomena, it functions on a purely statistical observation of correlations (untainted by any underlying logic) between data captured in an absolutely non-selective manner in a variety of heterogeneous contexts.

This translation and processing of reality reduced to data points and rendered predictable – in a data-rich environment such as ours, 'anything can be predicted' and 'crunching numbers' is 'the new way to be smart', Ian Ayres (2007) suggests – appears reassuring at different levels.

It appears reassuring especially at a time where narratives have become more than ever suspicious due to the experienced difficulty, in a multicultural, globalised society, to find common languages and emotional harmony with our fellow human beings. Rather than understanding the biographical trajectory and exotic world view of their foreign neighbour just moving in next door, Mister and Miss Anybody are interested in knowing in advance what risk the newcomer represents for their safety and tranquillity.

At the political level, the turn towards autonomic computing and a statistical governance of the 'real', for its orientation towards prediction, is a gesture that is both encouraged by and reinforcing a governmental rationality whose central figure is contingency and where prediction and avoidance of danger have replaced the identification and remediation to its causes. Suffice to observe the central themes of electoral campaigns in the Western world since 9/11 2001 to note that the ubiq-uitous figure of uncertainty has become so central that providing security through the anticipation of danger has eclipsed most competing political priorities at the governmental level. Globalisation seems to have ended the time – if there has ever been such a time – where governments could act towards an identifiable common good. How indeed could such a common good be identified in a global society such as ours, characterised above all by its cultural, economic, linguistic, religious frag-mentation, and by the palpable intensification of morally indefensible disparities in terms of health, wealth, and spending of scarce resources? Political, ecological, economic instability are the hallmarks of our 'risk society'. In a polity where the ubiquitous figures of contingency and risks have come to take the central space formerly (ideally) occupied by the figure of the common good, and where the pre-vention of insecurity, rather than the pursuit of any collectively identified common good has become the most important role of governments, where also, individuals are socialised through fear, a dominant phantasm is that autonomic computing, in allowing for the complex operations of data-mining and precise and dynamic profiling, will render the world and its inhabitants predictable. This provides the ideological background for enthusiastic support of any technology promising to help tame the chaos. The ubiquitous threat of virtual danger acts as a powerful incentive to eradicate whatever, in the human being, remains uncertain, virtual, potential.

Epistemically, I would suggest that it implements what Slavoj Zizek has identified as a shift from *modern* rationality to *post-modern* rationality, that is, the gradual replacement of 'transparency', allowing the understanding of profound mechanisms behind appearances ('transparency' as such was the privileged mode of modernity) by 'simulacra', the presentation of an impenetrable, but convivial surface. Ironically, according to the IBM vision of autonomic computing, the system will be 'transparent' not in the modern sense that it will allow the user to understand the deep mechanisms on which it functions, but in the sense of a total invisibility and imperceptibility of these mechanisms.

Such a post-modern rationality fits our post-modern governmentality: in the field of security, what disappears is the need to understand, explain and address the (too complex to grasp and address) causes of feared dangers. In the field of marketing, the logic relieves all actors from the burden of reflecting on possible discontinuities between (technologically persuaded) consumers' demand and their actual needs. The mobile, constantly reorganising and readjusting images of the 'real', highly relevant to private and public bureaucratic purposes, appear evaluated increasingly according to criteria of flexibility, speed and relevance, and decreasingly according to criteria of truth, objectivity, and justice. Isn't that the sign that, in the passionate pursuit of a phantasm of absolute predictability of events and persons, we are building simulacra, which, according to Baudrillard, are nothing but a 'copy without original' or a 'representation hiding the absence of reality (hyper-reality)'?[13]

Yet, by comparison with human observation, technologically intermediated observation may appear more 'objective': it appears to attest to a victory of rational analysis over deceptive human sensorial perceptions. Involving multimodal observation, these systems detect phenomena as they surface in physical and digital spaces, and privilege information ensuing from observation of, for example, the human body (making 'sense' from involuntary bodily movements, attitudes, physiological alterations). They follow the idea that, unlike human *persons*, human *bodies* do not lie.

Moreover, the substitution or addition of technological detection, classification, and forward-looking evaluation to human observation and judgement appears as a way to bypass ordinary biases and prejudices. Emotions are an essential element of human cognitive process, allowing individuals, unable to cope with the totality and the complexity of the world they live in, to prioritise certain (visual, sensorial, auditive, . . .) information, to ignore or forget a sufficient amount of the rest as to be able to keep reflecting and acting, which would be just impossible if they were constantly over-flown by information. The fact that some things are forgotten and others remembered is what gives human History a kind of normativity: ordinary lives are not inscribed in History. *Exemplar* existences and deeds are, and this filtering of the 'real' through human memory and historical inscription is how humans *transmit* normative evaluations from one generation to the other. Individual and collective human memory are of course not objective, but that lack of objectivity has proved absolutely necessary for the functioning of individuals, and for the organisation of societies. What all this suggests is that an intensive replacement of human observation, evaluation and prediction by autonomic processes may deprive us from some possibilities to still pose normative judgements at all.

That same danger of 'depoliticisation' and 'demoralisation' is carried by 'technological paternalism', and ramping in any technology designed for the purpose of rendering practically impossible behaviours, attitudes or actions that were previously 'simply' forbidden by morality or law (Spiekermann and Pallas 2006: 6–18).

Besides this important aspect, one may also note that the postulated reliability and impartiality of 'autonomically produced predictions' is vulnerable to a series of reasons identified by Gandy (2010) as

1   the possible inaccuracy of data used, or incorrectness of models or routines;
2   the fact that these are correlations-based systems possibly relying on categorical variables rather than causal inquiry;
3   the fact that these systems may produce 'rational' (facially non-biased) but 'unfair' results (disparately impacting in a disfavourable manner already vulnerable groups, in contradiction with common views of justice or fairness);
4   the lack of ground truths to evaluate the validity of detection mechanisms aimed at preventing certain behaviours to happen, as by hypothesis, these will not happen (the detection system can thus not be 'tested'), or aimed at complying with users' unexpressed needs or preferences (as these systems indeed *influence* these needs and preferences, according to the logic of 'dynamic nominalism' exposed by Ian Hacking, which I will describe later on).

Moreover, the type of knowledge so produced is in no way 'objective' in the sense that one has long been used to speak of the objectivity of scientific knowledge.

1. The information systems embedded in ambient intelligent systems are not intended to observe the unique complexity of each human being, but to *sort* individuals in a variety of heterogeneous categories for the purpose of predicting their willingness or need to buy specified commodities, their risks to fill claims with health and disability insurances, the danger they represent for themselves or for others, or other propensities that marketers, insurers, law enforcement officials and many others may find useful to have. Nikolas Rose summarised the phenomenon in these words:

> reduction of complexity by numbers can be neither ideologically nor theoretically innocent: hence the social enters the statistical through the 'interests' of those who undertake this task. The processes of simplification embody the

expectations and beliefs of the responsible technicians and officials. The discretion that they inevitably exercise is dissimulated by their claim that their expertise, whilst indispensable, is 'merely technical'.

(Rose 1999: 204)

2. The type of 'evidence' sustaining the knowledge so produced (the accuracy of profiles for example, and the reliability of predictions built thereon) is of a particular nature. Arguably, and quite counter-intuitively, I must concede, evidence, here has less to do with a process 'which consists in one thing pointing beyond itself' – the proof found in the understanding of the causes of phenomena – than with 'the rhetorical sense of vividness, a gesture which refers to the immediate appeal of the fact itself'.[14] This 'postmodernist' account of the 'real' has existentialist tonalities and reminds us of Jean-Paul Sartre's statement that 'L'évidence, c'est la présence pour la conscience de l'objet en personne (. . .) Une évidence, c'est une présence' (Sartre 1940: 201).

I am not the first scholar to notice that, nowadays, the present tends to prevail as the unique figure of authenticity: reality has become a concept entirely comprised in the present, the temporal mode of 'real-time': '[I]ndeed, one of the emerging constants in the theorization of futurity is that it is only the present which is *real* to us whereas the past and the future are only available to us through imagination and representation' (Brown 2003: 3–21). Whereas

> [w]e are used to thinking of modernity as defined in part by future-oriented ideals of progress, increasing technological control, and so on (. . .) modernity achieved its break with the past only by according the present the most profound normative and ontological privileges, and this privileging of the present eventually gave to modern man (. . .) as little reason to think of his society's future as he has to think of its past
>
> (Rubenfeld 2001:4)

> We have. . .confused Being with being-present. Nevertheless, the present *is not*; rather, it is pure becoming, always outside itself. It *is not*; but it acts. Its proper element is not being but the active or useful. The past, on the other hand, has ceased to act or be useful. But it has not ceased to be. Useless and inactive, impassive, it IS, in the full sense of the word: it is identical with being in itself.
>
> (Deleuze 1990: 55)

In the contemporarily dominant perspective, the current *presence* of things, rather than their spontaneous genealogy, is taken as a sign of their belonging to the domain of truth, authenticity or facts.

Indeed autonomically produced profiles render everything *actual*, present. They rely on digital, rather than human memory and therefore benefit of digital memories' virtually unlimited storage capacity in which, by default, everything is recorded, even the most trivial events, our most trivial, conscious or even unconscious gestures, and nothing is ever forgotten.[15] As a result, the construction of profiles, of the 'digital image' of individuals is, from the perspective of the later, a heteronomous construct, at odds with what, from an individual's point of view, counts as explanation, as agency, as causality, and at odds with how the individual could give an autobiographical account of himself as a being 'always over time', never fully comprised in the present, whose virtualities are never completely actualised, as a being which is a process rather than a substance.

3. These classifications made on the basis of statistical correlations have a feedback looping effect: Ian Hacking, coining on the occasion the concept of 'dynamic nominalism', explained that when people are taken as objects of scientific or bureaucratic inquiries for a variety of purposes going from controlling them to helping them, organising them, keeping them away from places. . .such classifications affect the people classified, and these affects on people, in turn, change the classification (Hacking 2007) in ways that are contingent on the type of finality and applications of the system (which is difficult to predict in advance). This results in the reinforcement and the 'viral propagation' of norms, of the criteria of normality and desirability against which individuals are being evaluated, with gratifications for compliant and sanctions for the others. Norms have always had a viral character though: Georges Canguilhem, already explained that the specificity of an object or fact said 'normal' by reference to either external or internal norms raises the possibility that it becomes itself taken as a reference for objects awaiting their characterisation as normal (Canguilhem 2005: 181). That viral character is only amplified by the intensification of profiling, and the ensuing phenomena of anticipative conformity, self-censorship or preferences falsification ensuing. My concern here is thus not merely the increased visibility of individuals (a traditional privacy issue). Rather, I am interested in the implications of the possibility that meaning is ascribed to even the most trivial and fugitive image, sound, movement transpiring from individual subjects. The processes through which meaning is produced follow a governmental rationality fitted to a world in which unpredictability and spontaneity (which are the virtual dimensions of human beings) are decreasingly tolerated, and where both the ubiquitous threat of the virtual danger (the risk) and the wish to adapt consumers to what the market has to offer rather than to adapt market offers to the genuine needs and preferences of consumers act as almost irresistible incentives to eradicate what, in the human being, remains uncertain, potential, *inactual*. But the actualisation of such a phantasm of a world liberated from contingency and unpredictability comes at an expensive price, as (or so I wish to argue), the potential, the *inactual*, are the modalities of individual and social existence which, through the conjugated powers of virtuality and utopia, provide the 'natural' preserves for individualisation and social change.

Technology, virtuality, utopia

Notwithstanding the unresolved(able) conceptual disagreements about its exact meaning, the 'virtual' has to do with the capacity human beings have to think and act by reference to something unfitted to the language and structure of the *actual* society, despite the inescapable fact that, as *subjects*, they are shaped by the *actual* language and structure. Virtuality, or our virtual dimension, has to do with the capacity we have to suspend any definition of ourselves, our capacity to 'think of ourselves beyond ourselves' in a cultivation of ecstasies or self-transcendence, self-overcoming or self-deconstruction.[16] This process of 'thinking ourselves beyond ourselves' is what is obviously involved in literary creativity. Michel Foucault explicitly acknowledged that this self-overcoming was among the fundamental reasons that made him write.

> Plus d'un, comme moi sans doute, écrivent pour n'avoir plus de visage. Ne me demandez pas qui je suis et ne me dites pas de rester le même: c'est une morale d'état-civil; elle régit nos papiers.
>
> (Foucault 1969: 28)

Virtuality is a concept as difficult to grasp as the concept of selfhood to which, I would argue, it is partially consubstantial, and of which it denotes the lack of substance and the essentially processual nature. What I wish to refer to is the kind of contingency and unpredictability of the form towards which individual personality may flourish. That virtuality is difficult to circumscribe. It is not a thing, but a process through which individuals become subjects, that is, tend towards an identity and personality that are never (entirely) pre-existing.[17]

> The virtual layer of a person's self, or whatever name one wishes to give to that fundamentally, and essentially, indefinable blind spot that a subject always consists of for himself, and which, one may argue, the right to privacy contributes to safeguard,[18] can arguably only be exercised in spaces or territories either mental or physical, which are not already saturated by *meaning*. A suggestive metaphor for this might be provided by the notion of 'junk DNA' present in our genomes but which does not seem to 'code' for any definite function. This 'junk DNA's crucial function, however, is to serve as a natural preserve for the evolution of our species.
>
> (Rouvroy 2008: 256)

Jean Baudrillard, posited a contemporary strategy of 'seduction' making individuals disappear into

> ever more sophisticated methods of biological and molecular control and retrieval of bodies', where 'the destiny of signs (. . .) is to be torn from their destination, deviated, displaced, diverted, recuperated, seduced. Everywhere one seeks to produce meaning, to make the world signify. We are not,

however, in danger of lacking meaning; quite the contrary, we are gorged with meaning and it is killing us.

(Baudrillard 1988: 63)

Disentangling the notion of the 'virtual' from contemporary fallacies, I wish to insist, as Gilles Deleuze and others did, that the virtual does possess a full reality, as virtual, and should not therefore be opposed, as it often is today, to the real.

> We opposed the virtual and the real: although it could not have been more precise before now, this terminology must be corrected. The virtual is opposed not to the real but to the actual. *The virtual is fully real in so far as it is virtual.* Exactly what Proust said of states of resonance must be said of the virtual: 'Real without being actual, ideal without being abstract'; and symbolic without being fictional. Indeed, the virtual must be defined as strictly a part of the object – as though the object had one part of itself in the virtual into which it plunged as though into an objective dimension.
>
> (Deleuze 1994: 108–12)

The virtual is as real as anything one can touch, except that it is not 'actual': it is a (potentially infinite) bundle of possibilities, living an existence which is *parallel* to the actual world of things and matters. It obviously has to do with utopia and may be supported by technology. Some artefacts indeed are *utopian*. So was the Internet in the early nineties. The cyber-space, at a time when the digital (ever since then called the *virtual*) and the physical spaces were radically separated from each other, was a place experienced as a true 'new world', 'a home for the Cyberpunks', 'and the whole Cyber Underground', a world that 'Hackers, Phreakers and other Cyberpunks began to rule (. . .) the way they like – No laws! No rules!'[19] This utopian space without location (Foucault 1984) was a place where new forms of thought, new forms of cooperation and interactions could be tasted fearlessly, as the radical deterritorialisation liberated from physical limitations as well as from all types of legal, parental, religious, and other authorities constraining actions and interactions in the physical world.

The freedom then experienced in the untamed cyber-space, *parallel* to and radically disconnected from the physical world did not survive to the gradual colonisation of the Internet by market logics, nor to the ascendency of search engine operators and other 'gatekeepers' equipped with unprecedented means to control users' experiences (deciding about the prioritisation of informational contents provided by search engines, designing and modifying the architecture of, and applications available in, social networks,. . .). From a space for liberated thought, communication, and experience, the Internet, set aside the survival of rare sub-spaces still devoted to creativity and experimental socialisation, has become a privileged space for *actualisation* of consumerism and conformism. This evolution may be described as an invasion of the digital space by the logics and authorities typical of the physical space, for which the digital space came to function as an amplifier. New 'persuasive technologies' are now emerging, notably from the Persuasive Technology Laboratory at Stanford University, with the explicit aim to shape people's opinions, preferences and attitudes, through technologies taking advantage of social dynamics and amplifying or weakening their effects and affects beyond the individual user's expectations and anticipations.[20] Technological architectures such as Facebook and other social networks, which people can access from anywhere at any time, from their laptop or their mobile phone, are well suited to the operation of such 'massively inter-personal persuasion'.[21] This is illustrative of how the same utopian technologies may turn into technologies of power. Altering both individual behaviour and motivation and the informational structures within which individuals behave, persuasive technologies appear to cumulate the strength of disciplinary and actuarial regimes.[22] The next stage in the development of our information society further blurs the separation between digital and physical realities. Imagining these detection, classification and forward-looking evaluation technologies, functioning, when useful, on massive inter-personal persuasion, becoming truly pervasive, ubiquitous and 'transparent' in the sense given by IBM to 'transparency' ('The system will perform its tasks and adapt to a user's needs without dragging the user into the intricacies of its workings'), is rather frightening. Projecting ourselves in such a dystopian picture allows, however, identifying what we are not ready to pay for living in a technologically tamed universe. Here we can at last pose a normative judgement. Against this dystopian projection, one may venture on the intimidating battlefield of values.

In a letter to Karl Jaspers, in which Hannah Arendt confessed herself how uncertain she was of what she wrote, she recounts:

> What radical evil is I don't know, but it seems to me it somehow has to do with the following phenomenon: making human beings as human beings superfluous – not using them as means, that does not infringe upon their humanity but merely upon their dignity of human beings, but rendering them superfluous despite their quality of being human. This happens as soon as unpredictability – which, in human beings, is the equivalent of spontaneity – is eliminated.
>
> (Arendt 1993: 166)

Infringing upon a being's humanity supposes the elimination of that being's unpredictability or spontaneity. It seems to me that this insistence on unpredictability and spontaneity as essential elements of what makes a being human is extremely important. It is anyway highly relevant for what I wish to say about the 'virtual' as an essential dimension of human beings: that virtuality – as unpredictability (even, up to a certain extent, for the subject himself), as spontaneity – appears as the hard core of what deserves the protection of the law not only, or even not necessarily, because failing to protect the unpredictability or spontaneity would infringe upon human dignity, but because it would directly infringe upon humanity itself, upon that which identifies beings as humans.

The 'virtual' layer was similarly identified by Gilles Deleuze as one of the imma-nent characteristics of human beings. I understand it as implying the fact for the human being, of never being fully comprised in the present, of always being 'over time'.

> In any case, the relation between the actual and the virtual is not that which one can establish between two actuals. The actuals imply already constituted individuals, and determinacies by ordinary points; whereas the relation between the actual and the virtual shape an individuation in action or a singularization by remarkable points to be determined in each case.
>
> (Deleuze 1996: 185, my translation)

This has much to do with what Michel Foucault called the process of subjectivation: the path through which individuals *become* subjects, a tension between the two poles constituted by, on the one hand, the self that I am, which I never completely possess – dependent as I am on interactions with others, on my capacity to give others and account of myself,[23] or to be interpellated by others and, on the other hand, the self I might, I may, I wish to become in the future, and which I cannot know in advance.

This unpredictability of human individuals, their spontaneity, is not merely worth preserving because of its contribution in making us human; it is also a necessary precondition to the vitality of society as a whole, which must remain open to changing its own basic rules and structures whenever these remain too far removed from the ideal of justice that people are able to imagine. Virtuality is to the indi-vidual human being what utopias are to societies. They are preserves for the 'flourishing of individual personality', and for fundamental changes in our social existence. This articulation between virtuality and utopia appears, in a subtle form, in Frederic Jameson's writings:

> Utopian form is itself a representational mediation on radical difference, rad-ical otherness, and on the systemic nature of social totality, to the point where one cannot imagine any fundamental change in our social existence which has not first thrown off Utopian visions like so many sparks from a comet. The fundamental dynamic of any Utopian politics (or of any political Utopianism) will therefore always lie in the dialectic of Identity and Difference, to the degree to which such a politics aims at imagining, and sometimes even at realizing, a system radically different from this one.
>
> (Jameson 2005: xii)

Autonomic computing, as a vision, is an ideological vision. Together with the increasingly 'intelligent' and 'autonomic' systems it is aimed to reinforce, it crystallises the dominant technological, economical and political projections or world views of our western time, but the specific representational regime it imple-ments may, this time, make it more difficult for individuals and groups to dissent.

Technologies blurring the separation between the 'physical' and the 'digital' on the one hand, and between the 'actual' and the 'virtual' on the other hand, unavoidably reconfigure human experience, setting new regimes of visibility and intelligibility and, as a consequence, impact on the terms through which one should think of power relations in society. All this attests to the actuality of Michel Foucault's writings on governmentality:

> As for all relations among men, many factors determine power. Yet ratio- nalization is also constantly working away at it. There are specific forms to such rationalization. It differs from the rationalization peculiar to economic processes, or to production and communication techniques; it differs from that of scientific discourse. The government of men by men – whether they form small or large groups, whether it is power exerted by men over women, or by adults over children, or by one class over another, or by a bureaucracy over a population – involves a certain type of rationality. It doesn't involve instrumental violence.
>
> (Foucault 1990: 84)

> Consequently, those who resist or rebel against a form of power cannot merely be content to denounce violence or criticise an institution. Nor is it enough to cast the blame on reason in general. What has to be questioned is the form of rationality at stake. The criticism of power yielded over the mentally sick or mad cannot be restricted to psychiatric institutions; nor can those questioning the power to punish be content with denouncing prisons as total institutions. The question is: how are such relations of power rationalised? Asking it is the only way to avoid other institutions, with the same objectives and the same effects, from taking their stead.
>
> (Foucault 1990: 85)[24]

Conclusion

The hypothesis I have sustained in this very tentative chapter is that the virtual dimension of individual human personality, which is constitutive of subjectivity itself, is incompatible with the actualisation – through technological or other means – of a depoliticised, statistical governmental rationality indifferent to the causes of phenomena and chiefly oriented towards the annihilation of contingency. I have also defended the idea that the eradication of virtuality (or subjectivity) is incom- patible with the emergence of utopias whereas utopias have a crucial role to play in sustaining the vitality of deliberative democracy. Thinking beyond oneself, individually, and beyond current societal configuration, collectively, are the indispensable reflexive capabilities allowing for individual self-determination and collective self-government. Autonomic computing may enhance or decrease these capabilities, depending on the governmental (or bureaucratic) rationality it is meant to serve.

Let us try not to miss the target here: choosing which technological evolution we wish to emerge in our life-world cannot be done without first having chosen which governmental rationality one wishes to have ruling our society. The debates must identify what the central figure of that governmental rationality should be.

Currently, the focus on contingency and risk minimisation has shadowed most other political goals. This obsession with contingency and virtual danger has also come to be raised by those who would like to see a 'right to security' explicitly acknowledged the status of a fundamental right, competing or even pre-empting the fundamental right to privacy, despite the obvious fact that the current regime of fundamental rights and liberties, as well as, for example, criminal law and the general principle of legal certainty (*sécurité juridique*) are already in place and ensuring a certain level of security. However, the rhetorical strategy followed by the advocates of an autonomous 'right to security' is aimed at justifying in advance systematic interferences by public authorities with the exercise by individuals of their right to privacy. The 'right to security' is in fact not a right providing individuals a legal basis to impose certain duties to act or certain abstentions on either the State, government officials, or fellow citizens. Rather, such a fundamental 'right to security' would amount to providing the State and government officials an advance and permanent justification for infringing upon fundamental individual rights and liberties, such as privacy, or freedom of expression, even in ordinary circumstances where no emergency threat would justify the temporary instauration of a 'state of exception', and even through means (including technological ones) which are not necessary in a democratic society in the interests of national security, public safety or the economic well-being of the country, for the prevention of disorder or crime, for the protection of health or morals, or for the protection of the rights and freedoms of others. Said otherwise, making a 'right to security' prevail over the right to privacy and other fundamental rights and liberties would radically jeopardise citizens' security (not to even speak of aliens), exposing them to the unconstrained and arbitrary surveillance, control, monitoring by law enforcement authorities and other officials. There are arguments to believe, indeed, that whatever technology will become available to intensify the scrutiny of individuals in all the dimensions of their life will find a market and be deployed, if only because these technological systems may be more cost-effective than the human workforce, and because information, especially when it allows forward-looking evaluation – or, even better, the orientation – of individual tastes, choices, behaviours, etc. has become both an invaluable asset for governments and commercial enterprises and the privileged yardstick of power to remote control the population.

There is now a crucial need to transcend the blinding choice between both the misled diabolization of technology and its uncritical endorsement. The issue about technology is to understand the kind of governmental rationality it may further sustain, to resist and enter the political struggle to counter it when it may lead us to a situation incompatible with the vitality deliberative democracy, and, together with other disciplines in human sciences, to cooperate with technology designers and politicians as to build an informational infrastructure allowing for the flourishing of human virtualities, even when these virtualities may give rise to radically new and unexpected individual and societal forms of existence. Throwing the dice is something that must, from time to time, be dared.

Notes

1 www.research.ibm.com/autonomic/overview/elements.html
2 I do not endorse the qualification of a turn towards autonomic computing as a 'paradigm shift' in a Kuhnian sense, the figure of the 'paradigm shift' is used by IBM essentially as an advertising metaphor.
3 See Sarat (1995: 15).
4 See Hayles (2006: 140): 'Enmeshed within this flow of data, human behavior is increasingly integrated with the technological nonconscious through somatic responses, haptic feedback, gestural interactions, and a wide variety of other cognitive activities that are habitual and repetitive and that therefore fall below the threshold of conscious awareness. Mediating between these habits and the intelligent machines that entrain them are layers of code. Code, then, affects both linguistic and nonlinguistic human behavior. Just as code is at once a language system and an agent commanding the computer's performances, so it interacts with and influences human agency expressed somatically, implemented for example through habits and postures. Because of its cognitive power, code is uniquely suited to perform this mediating role across the entire spectrum of the extended human cognitive system. Through this multilayered addressing, code becomes a powerful resource through which new communication channels can be opened between conscious, unconscious, and nonconscious human cognitions.'
5 See (Fuller 1994: 741).
6 Alford and Friedland (1985: 165): 'the legitimating principles for different institutions, and their conflicts constitute the society and help explain its structure and changes.'
7 I opted for not considering autonomic computing independent from the applications in which this 'new paradigm' may be involved.
8 I am indebted to Massimo Durante for this emphasis on identity as something that may only be 'said' by a third person.
9 Dean (1999: 2): 'The term *governmentality* seeks to distinguish the particular mentalities, arts and regimes of government and administration that have emerged since "early modern" Europe, while the term *government* is used as a more general term for any calculated direction of human conduct. Typical of his flair for a catchy and perspicacious phrase, Foucault redefined "government" in a fashion compatible with its sixteenth- and seventeenth-century uses as the "conduct of conduct", i.e. as any more or less calculated means of the direction of how we behave and act.'
10 Ontogeny is the study of the spontaneous process through which the same living organism undergoes structural changes while remaining organised in a way that ensures that it remains the 'same' organism. See Maturana and Varela (1986).
11 This has probably never been more obvious than today, given the shifts experienced in the modes of knowledge production: rather than merely discovering nature's secrets, scientists and technology designers increasingly produce knowledge and technological devices in the context of problems and applications defined by funding agencies concerned with specific agenda (such as European competitivity, bureaucratic efficiency or business profitability).
12 Castel (1984: 237–38): 'problematisation is not the representation of a pre-existing object, or the creation through discourse of an object that does not exist. It is the totality of discursive and non-discursive practices that brings something into the play of truth and falsehood and sets it up as an object for the mind.'

13   See, however, Deleuze (1969: 302–3): 'Le simulacre n'est pas une copie dégradée, il
     recèle une puissance positive qui nie *et l'original et la copie, et le modèle et la
     reproduction.*'
14   The distinction is recalled by Shaffer (1992: 328), referring, for the description of the
     first notion of evidence, to Hacking (1975).
15   On this issue, see our previous work, Rouvroy (2009).
16   Scott (1992: 106–7): 'our subjectivity has been formed in a process of "subjectivation"
     in which we have come to relate to ourselves by values which overlook our own
     fragmented histories and thereby carry a largely unconscious inclination towards
     totalization and fascism. Instead of thinking of him as uncommitted and, by thinking
     this way, holding our thought within the framework of the committed or uncommitted
     subject, one can think of *attunement* to something that is not fixable within the
     boundaries of subjects and representations. I do not know how to speak of something
     so out of the bounds of representation and subjectivity.'
17   In that sense, the contemporary injunction to 'be oneself' is nonsensical.
18   At least according to legal scholars (e.g. Rubenfeld 1989) and courts (e.g. the German
     Supreme Court) acknowledging an anti-totalitarian concept of privacy that is protective
     of the individual's right to freely develop his/her personality, and the correlative need
     for some level of 'seclusion' as a precondition to the flourishing of individual personality.
19   Mad Maniac (1996), http://project.cyberpunk.ru/idb/history.html.
20   See B.J. Fogg (2008: 23–34).
21   About the Obama campaign on Facebook, read www.facebook.com/note.php?note_id=
     46049223571&id=8417788415&index=0).
22   See Simon (1988: 773): 'Disciplinary practices focus on the distribution of a behavior
     within a limited population (a factory workforce, prison inmates, school children, etc.).
     This distribution is around a norm, and power operates with the goal of closing the gap,
     narrowing the deviation, and moving subjects toward uniformity (workers are to be
     made more efficient and reliable, prisoners more docile, school children more attentive
     and respectful). Actuarial practices seek instead to map out the distribution and arrange
     strategies to maximise the efficiency of the population as it stands. Rather than seeking
     to change people ("normalize them", in Foucault's apt phrase), an actuarial regime seeks
     to manage them in place.'
23   Butler (2005: 136): '[T]o be undone by another is a primary necessity, and anguish to
     be sure, but also a chance – to be addressed, claimed, bound to what is not me, but also
     to be moved, to be prompted to act, to address myself elsewhere, and so to vacate the
     self-sufficient "I" as a kind of possession. If we speak and try to give an account from
     this place, we will not be irresponsible, or, if we are, we will surely be forgiven.'
24   See also Curtis (2002: 505–33).

References

Akrich, M. (1992) 'The De-Scription of Technological Objects', in. Bijker, W.E. and Law,
     J. (eds), *Shaping Technology / Building Society*, Cambridge: London: MIT Press.
Alford, R.R. and Friedland, R. (1985) *Powers of Theory: Capitalism, the State, and
     Democracy*, Cambridge: Cambridge University Press.
Arendt, H. (1993) *Correspondence Hannah Arendt – Karl Jaspers 1926–1969*, San Diego;
     New York, London: Harvest Books.
Aristotle (1979[350BC]) *Generation of Animals*, Harvard: Harvard University Press, Loeb
     (Loeb Classical Library).
Ayres, I. (2007) *Super Crunchers: Why Thinking-by-Numbers is the New Way to be Smart
     (using data to make better predictions)*, New York: Bantam.

Baudrillard, J. (1988) *The Ecstasy of Communication*, New York: Semiotext(e).

Brown, N. (2003) 'Hope against Hype – Accountability in Biopasts, Present and Futures', (16) *Science Studies*, 2: 3–21.

Butler, J. (2005) *Giving an Account of Oneself*, Fordham University Press.

Canguilhem, G. (2005 [1966]) *Le normal et le pathologique*, Paris: PUF.

Castel, R. (1984) 'Problematization as a Mode of Reading History' in Goldstein, J. (ed.), *Foucault and the Writing of History*, Oxford: Blackwell: 237–52.

Curtis, B. (2002) 'Foucault on Governmentality and Population: The Impossible Discovery', (27) *Canadian Journal of Sociology*, 4: 505–33.

Dean, M. (1999) *Governmentality. Power and Rule in Modern Society*, London, Thousand Oaks, Calif.: Sage.

Deleuze, G. (1969) *La logique du sens*, Paris: Minuit.

—— (1990) *Bergsonism*, New York: Zone Books.

—— (1994) *Difference and Repetition*, translated by Paul Patton, New York: Columbia University Press.

—— (1996) 'L'actuel et le virtuel' in *Dialogues*, Paris: Flammarion: 179–81

Finlayson, A. (2006) '"What's the Problem?": Political Theory, Rhetoric and Problem-Setting', (9) *Critical Review of International Social and Political Philosophy*, 4: 541–57.

Fogg, B.J. (2008) 'Mass Interpersonal Persuasion: An Early View of a new Phenomenon', in *Persuasive Technology*, Harri Oinas-kukkonen, Per Hasle, Marja Harjumaa. . .[et al.] (eds), Berlin, New York: Springer: 23–34.

Foucault, M. (1969) *L'archéologie du savoir*, Paris: Gallimard.

—— (1980) *Power/Knowledge: Selected Interviews and Other Writings, 1972–1977*, New York: Pantheon Books.

—— (1982) 'The Subject and Power', (8) *Critical Inquiry*, 4: 777–95.

Foucault, M. (1984) 'Des espaces autres', *Architecture / Mouvement / Continuité*, 5: 46–49. English transl. by May Miskowiec, 'Of Other Spaces: Heterotopias' available at http://foucault.info/documents/heteroTopia/foucault.heteroTopia.en.html (last visited 3 July 2010)

—— (1990) 'Politics and Reason', in Foucault, M. (ed.), *Politics, Philosophy, Culture: Interviews and Other Writings, 1977–1984*, London: Routledge: 57–85.

—— (1991a) 'Governmentality', in Burchell, G. Gordon, C. and Miller, P. (eds), *The Foucault Effect: Studies of Governmentality*, Chicago: University of Chicago Press: 87–104.

—— (1991b) 'Polemics, Politics, and Problematisations: an Interview', in Rabinow, D. (ed.), *The Foucault Reader,* Harmondsworth: Penguin.

—— (1994) 'La gouvernementalité', in Foucault, M. *Dits et écrits*, Paris: Gallimard: 635–57.

Fuller, S. (1994) 'Making Agency Count. A Brief Forray into the Foundations of Social Theory', (37) *American Behavioural Scientist*, 6: 741–53.

Gandy, O. H. (2010) 'Engaging Rational Discrimination: Exploring Reasons for Placing Regulatory Constraints on Decision Support Systems', (12) *Ethics and Information Technology*, 1: 29–42.

Gordon, C. (1991) 'Governmental Rationality: An Introduction', in Burchell, G., Gordon, C. and Miller, P. (eds), *The Foucault Effect: Studies in Governmentality*, Chicago: Chicago University Press: 1–52.

Hacking, I. (1975) *The Emergence of Probability: A Philosophical Study of Early Ideas about Probability, Induction and Statistical Inference*, New York and London: Cambridge University Press.

—— (2007) 'Making Up People', (26) *London Review of Books*, 16, 17 August 2007.

Hayles, K. (2006) 'Traumas of Code', (33) *Critical Inquiry*, 1: 140.

Jameson, F. (2005) *Archaeology of the Future (Poetics of Social Forms)*, London, New York: Verso.

Latour, B. (2005) *Nous n'avons jamais été modernes: Essai d'anthropologie symétrique*, Paris: La Découverte.

Mad Maniac (1996) *Mad Maniac*, available at http://project.cyberpunk.ru/idb/history.html (last visited 12 November 2010).

Maturana, H. R. and Varela, F. (1986) *The Tree of Knowledge: The Biological Roots of Human Understanding*, Boston: Shambhala.

Rose, N. (1999) *Powers of Freedom*, Cambridge: Cambridge University Press.

Rouvroy, A. (2008) *Human Genes and Neoliberal Governance: A Foucauldian Critique*, Abingdon and New York: Routledge-Cavendish (GlassHouse Books).

—— (2009) 'Réinventer l'art d'oublier et de se faire oublier dans la société de l'information?', in Lacour, S. (ed.), *La sécurité de l'individu numérisé*, Paris: L'Harmattan.

Rubenfeld, J. (1989) 'The Right of *Privacy*' (102) *Harvard Law Review*, 4: 737–807.

—— (2001) *Freedom and Time: A Theory of Constitutional Self-Government*, New Haven: Yale University Press.

Sarat, A. (1995) 'A Prophecy of Possibility: Metaphorical Explorations of Postmodern Legal Subjectivity', (29) *Law and Society Review*, 4: 615–30.

Sartre, J.-P. (1940) *L'imaginaire*, Paris: Gallimard.

Scott, Ch. E. (1992) 'Foucault, Ethics, and the Fragmented Subject', *Research in Phenomenology*, 2 (1): 104–37.

Shaffer, S. (1992) 'Self Evidence', (18) *Critical Inquiry*, 2: 327–62.

Simon, J. (1988) 'The Ideological Effects of Actuarial Practices', (22) *Law and Society Review*, 4: 771–800.

Skouteris, V. (2007) 'Statistical Societies of Interchangeable Lives', (15) *Law and Critique*, 2004, 2: 119–38.

Spiekermann, S. and Pallas, F. (2006) 'Technology Paternalism – Wider Implications of Ubiquitous Computing', (4) *Poiesis & Praxis: International Journal of Technology Assessment and Ethics of Science*, 1: 6–18.

Stengers, I. (2002) *Sciences et pouvoirs. La démocratie face à la technoscience*, Paris: La Découverte.

Verbeek, P.-P. (2008) 'Cyborg Intentionality: Rethinking the Phenomenology of Human-Technology Relations', (7) *Phenomenology and Cognitive Science*, 3: 387–95.

Welch, M. (2008) 'Foucault in a Post-9/11 World: Excursions into Security, Territory, Population', (4) *Carceral Notebooks*, available at www.thecarceral.org/journal-vol4.html (last visited 12 November 2010).

Winner, L. (1986) 'Do artifacts have politics?' in Winner, L., *The Whale and The Reactor: A Search for Limits in an Age of High Technology*, Chicago: University of Chicago Press.

Chapter 8

Autonomic and autonomous 'thinking'

Preconditions for criminal accountability

Mireille Hildebrandt

Introduction

Cognitive psychology suggests that unconscious 'thought' is capable of complex analyses, way beyond the capability of the conscious mind. In fact, many cognitive scientists claim that most – if not all – of our behaviour is the result of 'the adaptive unconscious' without which we would not be able to function at all. Explaining or justifying our actions depends on being consciously aware of what motivated these actions. If, however, the 'causes' of our behaviour are not accessible to the conscious mind, the 'reasons' we give may be qualified as a comfortable illusion. The difference between autonomic computing and human behaviour, in that case, is not obvious.

This would either mean that attributing criminal liability to autonomic computing systems should not be a problem or it would mean that attributing criminal liability to human beings makes no sense anyway.[1] In this paper I will challenge such a position by elaborating the distinction between autonomic and autonomous 'thought', claiming its relevance for criminal accountability. I will build on recent findings of cognitive psychology about 'The New Unconscious', relating this to Judith Butler's exploration of the constitutive opacity of the self, explored in the traditions of Hegel, Nietzsche, Freud and Foucault.

My argument will be that the interplay between autonomic and conscious thought is the precondition for autonomous action, constituting the possibility to blame a person for wrongful action. This is relevant to the extent that autonomic computing environments can manipulate our autonomic behaviours in ways that do not reach the threshold of consciousness, thus easily manipulating us into being nice and decent (or horrible and dangerous) individuals without us having a clue as to why we act as we do.

Autonomic computing and the autonomic nervous system[2]

Autonomic computing

In 2001 IBM launched a project on 'autonomic computing'. IBM claims autonomic computing to be the only answer to the increased complexity that will arise from the expansion of interconnected networked environments. If we imagine the realisation of Embedded Wireless Sensor Networks,[3] integrated into an Internet of Things (ITU 2005) we come close to what has been called a vision of Ambient Intelligence (AmI) (Aarts and Marzano 2003, ISTAG 2001). According to IBM the ensuing complexity will soon be beyond repair (Kephart and Chess 2003: 41), requiring a new focus for research that should target the development of computer systems capable of self-management, self-configuration, self-optimisation, self-healing and self-protection. The claim is that an environment that is capable of anticipating our inferred wishes in real time, always on the alert for changing circumstances, cannot be realised without a measure of autonomic computing.

On the IBM website, dedicated to autonomic computing, we find the explanation for the use of the concept 'autonomic':

> The most direct inspiration for this functionality [real time adaptation of the environment, MH] that exists today is the autonomic function of the human central nervous system. Autonomic controls use motor neurons to send indirect messages to organs at a sub-conscious level. These messages regulate temperature, breathing, and heart rate without conscious thought. The implications for computing are immediately evident; a network of organized, "smart" computing components that give us what we need, when we need it, without a conscious mental or even physical effort.

This is an interesting proposition, using a biological metaphor. The decisions made by our autonomic nervous system are subconscious in a non-Freudian sense; we simply have no access to them and even if we did, the language in which the various parts of our organic system communicate is not symbolic or meaningful in a way that our conscious mind would be capable of understanding. This is not to say that the unconscious is not a part of our self: if the system breaks down this is the end of us. We do not only *have* an autonomous nervous system but we actually *are* our autonomous nervous system. This does not imply that this is all there is to us, but realising its constitutive role should bring home some of the challenges presented by visions like AmI and by autonomic computing as their precondition.

Profiling and pattern recognition

The idea of autonomic computing is that our external environment begins to cater to our needs in a similar vein as our internal environment. This requires continuous profiling of our every movement, biological states, interactions, moods and responses to what happens in the environment. Instead of waiting for our deliberate input on how we want our coffee, what room temperature we like, which music fits our mood, which is the best moment to wake up, how we prioritise and respond to incoming information, the environment profiles our keystroke behaviour and correlates the patterns it exhibits to our health, our mood, our productivity and – for instance to the moment we need a shot of strong black or mild and milky coffee. This is just one example. The range of patterns that can be disclosed by profiling technologies seems unlimited. A combination of facial expressions, like the blink of an eye, yawning and the size of the iris, can correlate with a driver's fatigue (Jin et al. 2007), enabling a smart car to decide when to prohibit speeding or driving per se. Ambient Intelligence implies persistent, pervasive and ubiquitous monitoring and this has given rise to a state of alarm with privacy activists. The fear is that this monitoring provides extensive traces of what we did, when, where, how, and with whom. I think this fear is partly unjustified, because the sheer amount of trivial data collected and stored will turn the information they could present into noise. Once collected and stored these data become decontextualised, devoid of meaning, empty digital signs.

My own fear, however, is a different one. Either all these data are lost because we have no means to retrieve them, to recontextualise them, to provide meaning and use them, in which case the whole enterprise is a waste of time, money and energy, *or* we develop techniques to trace and correlate the data, to detect patterns invisible to the naked eye, to recombine discrete data into new contexts and to anticipate futures on the basis of patterns found in past behaviours (Custers 2004, Fayyad et al. 1996, Hildebrandt 2008a and 2008b, Hildebrandt and Gutwirth 2008, Kallinikos 2006 and 2008). Wasting time, money and energy is a serious matter, but what concerns me here is the other possibility.

Imagine a smart energy meter that provides information on how much energy you are using at any moment in time in pence per hour, tonnes of CO_2 or kWh.[4] This could allow you to change your energy spending habits in simple ways, since you have the relevant feedback to make effective decisions. However, if this information is collected and stored with the owner of the infrastructure and disclosed to government officials or sold to commercial enterprises the data could be mined for significant patterns that reveal intimate details of your lifestyle. More interestingly, such data can be aggregated on a large scale, enabling data mining operations on the data of an entire population. In that case, if data on energy usage per quarter of an hour are correlated with data on health, employment, travels, shopping, political and religious affiliation, refined group profiles will emerge that correlate specific patterns of energy usage with a specific earning capacity, health status, shopping preferences, etc. Such profiles, even though they are based on statistical inferences and merely present probabilities, can be applied if they match with your data, and this application will impact on your life.[5] Autonomic profiling, as a precondition for smart environments, means that unprecedented amounts of data will be mined for relevant patterns while these patterns will be used to anticipate your preferences and to proactively adapt the environment to fit your priorities. This is what will give smart environments a competitive advantage over others that have not succeeded in accommodating to its users. This is how smart environments may develop a new cognitive power that could eventually match the cognitive power of living creatures, whose continued existence in a changing environment depends on adequate pattern recognition.

The idea that profiling or pattern recognition is the hallmark of life, of keeping in touch with and attuning one's responses to one's ecological niche, can be traced in the work of Maturana and Varela (1991, where it could be defined as structural coupling between an autopoietic system and its environment), James Gibson (1986, where it could be defined as locating the affordances of the environment by an organism) but also of Herbert Simon who defined intuition as subconscious pattern recognition that can – in principle – be performed by intelligent machines (Franz and Simon 2003). This chapter aims to prepare the comparison of unconscious pattern recognition by individual humans with machine profiling, in order to investigate to what extent autonomic computing makes a difference to human autonomy and thereby criminal liability. This is highly relevant for the legal framework of constitutional democracy in as far as this both assumes and produces a measure of human autonomy (Hildebrandt 2008b).

The opacity of the self: autonomic and autonomous thought

The wrong question: do reasons causes actions?

In *Ethical Know-How. Action, Wisdom, and Cognition* Francisco Varela (1999: ix) undertakes 'an understanding of ethics in a nonmoralistic framework'. He opens his first chapter with the proposition that 'Ethics is closer to wisdom than to reason, closer to understanding what is good than to correctly adjudicating particular situations' (Varela 1999: 3). He refers to Charles Taylor's distinction between a moral philosophy that builds on the Kantian tradition of moral judgement and an ethics of situatedness that builds on the tradition of Hegel. He explains the difference by stating that (Varela 1999: 4):

> As a first approximation, let me say that a wise (or virtuous) person is one who knows what is good and spontaneously does it. It is this immediacy of perception and action which we want to examine critically. This approach stands in stark contrast to the usual way of investigating ethical behavior, which begins by analyzing the intentional content of an act and ends by evaluating the rationality of particular moral judgements.

The question that comes to mind is how we can be sure that a person is right about what she thinks to be good and a moral philosopher would probably retort that this is exactly what moral philosophy is about. This indicates a division of labour: those interested in embodied and situated action investigate how a person implements what she knows to be good, while those interested in criteria to decide which behaviour is right investigate how we can develop such criteria. However, to the extent that such a division of labour implies that knowing precedes acting, it reduces action to the implementation of representational knowledge about the right way to act.[6] In speaking of 'the immediacy of perception and action' and 'wisdom rather than reason' Varela seems to suggest that ethical knowledge is a matter of perception instead of a result of correct reasoning. His usage of the term 'immediacy' of both perception and action seems to indicate that in perceiving a situation we intuit how to act while performing the action, and the usage of the term 'spontaneously' indicates that the action is not premeditated by a conscious thought. One could say that the action is not 'caused' by our conscious will.

This is a controversial proposition that flies in the face of the idea that an action is 'good' or 'right' if we can give good reasons for it and have performed it with those reasons in mind. Interestingly Varela's position has some affinity with recent findings in neuroscience and cognitive psychology 'demonstrating' that the whole idea that our intentions 'cause' our behaviour is an illusion. Following Libet's experiments in the beginning of the 80s several authors have defended the position that conscious intention *does not precede but actually follows* motor action (Haggard and Libet 2001, Wegner 2002). Since to be a cause something must precede its consequence, conscious thought cannot be the cause of our actions. It rather seems to be a by-product of actions, providing us with a sense of control, while in fact the causes reside in complex brain mechanics to which we have no conscious access whatsoever.

The problem with Libet's experiments is that they concern simple motor actions, like the movement of a hand. It is unclear what conclusions can be drawn from these kinds of controlled laboratory experiments for more complex behaviours, for instance for those that are the result of formal learning processes. Another question raised by the experiments is how the subconscious brain mechanics that are supposed to 'cause' the behaviour have been inscribed in the brain. If this was not part of our genetic make-up, this inscription must be the result of a learning process that may have involved conscious thought, especially in the case of more complex behaviours that depend on linguistic interaction. The fact that much of our behaviour is automated does not imply that we have not consciously initiated this automation at some point in the past.

To complicate the issue further, Dijksterhuis (2006) discusses experiments that demonstrate the capacity of *unconscious thoughts* to deal with complex decision-making. His experiments 'demonstrate' that allowing for unconscious thought processes provides for markedly better decisions – in the case of complex problems – than either conscious thought or immediate decision-making. This seems to require a qualification of Varela's idea of immediate spontaneous action as a superior type of behaviour, reflecting knowledge of what is good. Well-tuned spontaneous action may require a measure of time during which the action is prepared at a level of subconscious thought processes, capable of more refined and complex 'processing' of information than conscious thought will ever be.

Varela's position can also be linked to the writings of Goldberg (2006) on wisdom as sedimented pattern recognition that is typical for the brain of an experienced person, providing a reliable resource of automated (re)actions. The subtitle of his *The Wisdom Paradox* (2006) is *How Your Mind Can Grow Stronger As Your Brains Grow Older*. Pattern recognition is supposedly sedimented in the left hemisphere of the brain, being the physical correlate of what we tend to call our intuitions about how to understand a situation and of how to act in it. According to Goldberg this intuition is not pre- or nonanalytic but postanalytic, because it is the result of a learning process that can be traced in the brain as a matter of inscribed pattern recognition. This concurs with Gigerenzer's nuanced discussion of gut feelings as a form of unconscious intelligence (Gigerenzer 2007), providing the background of Gladwell's popular exaltation of gut feelings in his bestselling *Blink* (2005). What makes Gigerenzer more interesting than Gladwell is that he aims to discuss *when* we can trust our gut feelings, meaning that we cannot take our intuitions for granted as being right or good. Indeed, Varela did not write that a wise person is one who spontaneously acts in accordance with whatever she feels is good; he specifically wrote that a wise person is one who knows what is good and does it.

We conclude that whereas a Cartesian perspective invites the idea that thoughts are the causes of actions, cognitive science suggests that in fact actions 'cause' our thoughts even while having us believe that we are in control. Actions themselves are 'caused' by subconscious brain processes to which we have no conceptual access. So, we have thoughts, unconscious thoughts, intentions, behaviours and actions and we have the personal experience of control over our actions and – just like in the case of our memory (Loftus and Ketcham 1994) – psychology is demonstrating that the control we think we have over our behaviour may be an illusion.

From a non-Cartesian perspective this is not necessarily a surprise. The idea that autonomous thought presumes an individual will that is the cause of movements and more complex behaviour but has not been caused itself because it falls within the domain of the untouchable *res cogitans*, is a problematic idea if not the engine of a series of illusions about material passivity and causality on the one hand, and mental activity and freedom on the other hand. In trying to solve the problems generated by the separation of a *res extensa* and a *res cogitans* empiricist and rationalist traditions end up with the mentalistic picture of an uncaused will that causes our behaviour. That science now proves this to be an utterly inadequate model for understanding the complex relations between thinking and acting, actually confirms longstanding suspicions of the self as an independent subject whose inner freedom can be taken for granted. Nietzsche, Freud and Foucault seem to line up here with the latest findings of the cognitive sciences – *bien étonnés de se trouver ensemble*.

This, however, does not imply that these findings cannot clarify the relationship between automated behaviours, intentional action and the meaning of human freedom. In what follows I will first explore the nature of autonomic thought, by discussing some salient experiments performed within the realm of the cognitive sciences. After that I will try to come to terms with the idea of autonomous action as the hallmark of human freedom, safely rooted in a constitutive opacity that grounds the self in the abyss of its own inaccessible beginnings. Returning to Varela I will argue for an enacted view of ethical perception, which allows me to understand autonomic thought as a preconditional dimension of autonomous thought. This paves the way for the articulation of the right question: to what extent is autonomic thought compatible with autonomous thought and intentional action? This question is of crucial importance for the attribution of criminal liability.

Autonomic thought: pre-, non- or postanalytic?

Referring to Herbert Simon's work on pattern recognition Goldberg (2006: 104) understands intuition as a postanalytic competence, the result of a learning process that enables automatic pattern recognition and thus affords what he calls a 'mental economy'.[7] Instead of reinventing the wheel by consciously deliberating on how to respond to a specific situation, the mind recognises types of situations and automatically fits them with previous responses. It should be obvious that this process is not mechanical in any simple way and we may readily assume that this process of pattern recognition is in fact too complex to be represented in the conscious mind, since it has to take into account a myriad of interrelated features of an environment that is never the same as when the pattern (the type of situation) was first encountered. Also, the response to a similar type of pattern cannot be a simply copy of a previous response, so below the threshold of consciousness an unconceivable amount of work is done to attune our responses to whatever challenges we face. Gigerenzer (2007: 18–19) explains that this does not mean that the brain solves a complex problem with a complex strategy, following logical principles. The unconscious brain does not specify all consequences for each action, weighing them carefully and adding up the numbers in order to choose the one with the highest value or utility. This kind of deliberation would be more typical for the conscious mind, at least for the conscious mind of someone trying to be rational about things. The unconscious brain – according to Gigerenzer – does not function like a calculating machine, 'rather, nature gives humans a capability, and extended practice turns it into a capacity' (idem: 18). He speaks of the unconscious as '*knowing, without thinking*, which rule is likely to work in which situation' (idem: 19, my emphasis).

Other authors, however, claim that *the unconscious does think, but we just don't know it*. In his 'A Theory of Unconscious Thought' Dijksterhuis and Nordgren (2006) describe an interesting set of experiments. For instance, participants were given information about four apartments, described in terms of 12 different features. One of the apartments was significantly more attractive than the others, while another one was significantly less attractive than the others. The 48 bits of information about the apartments were presented to the participants in a random manner, confronting them with a 'daunting amount of information' (idem: 96). After that they were asked to evaluate the apartments. One-third of the group had to perform this evaluation immediately after the information was provided, leaving no time to think things through. Another third of the group were given three minutes to think about the information before their evaluation. The last third were told they would be asked about the evaluation later; this group was distracted for three minutes before their evaluation. Surprisingly, perhaps, those who had time to unconsciously 'think' about the assignment did best in terms of coining the most and least attractive apartments as such.

In another experiment Dijksterhuis and Van Olden (2006) gave participants a choice between five different art posters. One-third of the group were presented with all five posters at the same time and had to decide immediately; one-third was presented with each poster separately for 90 seconds and were asked to review each of them and to write down their reasons for liking or not liking a poster and to carefully analyse their preferences; a last group was presented with all five posters and told that at a later moment they would be asked to choose the one they liked best (they were then distracted with a task of solving anagrams during 450 seconds). After the participants made their choice they answered some questions and – to their surprise – were given the poster of their preference. This time the researchers had no *ex ante* qualification of which was the best option. Instead, they asked participants at a later point in time (3–5 weeks after the experiment) to indicate their satisfaction with the choice they had made, the regrets they might have developed about their choice and the amount of Euros for which they were willing to sell the poster. The final outcome of the experiment was that those who had time for unconscious thinking were more satisfied with their choice, had less regrets and asked a higher price for an eventual sale of their poster. Dijksterhuis and Van Olden (2006: 630) conclude that:

> People who were given the opportunity to think unconsciously about choices made superior decisions relative to those who thought consciously or who did not think at all.

It is not very difficult to criticise this conclusion on the basis that it assumes that – what the authors call – a higher 'post-choice satisfaction' is an indication of a superior decision. There may be other reasons or causes that explain the correlation between having some time between seeing the options and deciding while being distracted and being more happy with one's choice after a number of weeks. For instance, one could counter that not having consciously scrutinised the posters makes the choice less vulnerable to doubt, assuming that participants have not given their choice much thought after the experiment was over. In that case the higher 'post-choice satisfaction' doesn't mean very much. However, this does not explain the difference between those who chose immediately and those who had some time to unconsciously prepare the choice. I will not continue to question the conclusions at his point, because I want to take up the challenge posed by similar findings about the workings of the unconscious mind.

It seems evident that the conscious mind does well in solving relatively simple problems that do not require taking into account too many interrelated factors, which would require exponential processing powers compared to those available to the conscious mind. Such exponential processing powers seem to be a capability of the unconscious mind but we have no conscious access to them. 'They' (these brain processes) 'think for us' and we are bound to trust their outcome up to the point where we begin to reflect upon them and *change our mind*. This reflection may lead us to change our courses of action and when this implies changing our habits it will involve ingraining or inscribing new types of pattern recognition into the unconscious mind. Once these patterns are wired into the wetware of our mind, autonomic thinking – and acting – can take over and we can direct our conscious attention elsewhere, trusting that our unconscious awareness takes care of the rest. The conscious mind thus allows for postanalytic autonomic thinking,[8] but we can assume that our unconscious mind develops most of its autonomic behaviours without conscious input. This means that much autonomic behaviour is learned but nevertheless pre- or nonanalytic. Indeed, most of our self seems to evolve below the threshold of consciousness, taking care of our physical and emotional equilibria in the flux of events amidst which we live and sustain our self. The opportunities for conscious reflection are limited and their function seems to be just this: directing conscious attention to an issue that has emerged into consciousness, allowing for some direction over how to cope with it, providing an occasion to attune one's autonomic responses to goals that one can consciously articulate.

Autonomic 'thinking' then, must be a dynamic *mélange* of pre-, non- and post-analytic 'thinking', with most of these processes being inaccessible for conscious reflection (nonanalytic) even if some are the result of conscious interventions (postanalytic) and still others are on the verge of emerging in the conscious mind, asking for our focused attention (preanalytic).

Autonomous thought and the constitutive opacity of the self

The question that surges into my consciousness at this point, is how to understand the idea of autonomous thought in the light of the constitutive opacity of the self under discussion, and the related question of what room is left for what we like to think of as human freedom. To explore this issue we need to leave the realm of motor action and relatively simple consumer choice, taking into account the complexity of contemporary society and of the technological infrastructure that affords such complexity.[9] I will initiate this further exploration with a substantive quotation taken from Judith Butler's *Giving an Account of Oneself* (2005: 65):

> To say, as some do, that the self must be narrated, that only the narrated self can be intelligible and survive, is to say that we cannot survive with an unconscious. It is to say, in effect, that the unconscious threatens us with an insupportable unintelligibility, and for that reason we must oppose it. The 'I'

who makes such an utterance will surely, in one form or another, be besieged by what it disavows. An 'I' who takes this stand – and it is a stand, it must be a stand, an upright, wakeful, knowing stand – believes that it survives without the unconscious. Or, if it accepts the unconscious, this 'I' accepts is as a possession, in the belief that the unconscious can be fully and exhaustively translated into what is conscious. It is easy to see that this is a defended stance, but it remains to be seen in what this particular defense consists. It is, after all, the stand that many make against psychoanalysis itself. In the language that articulates opposition to a non-narrativizable beginning resides the fear that the absence of narrative will spell a certain threat, a threat to life, and will pose the risk, if not the certainty, of a certain kind of death, the death of a subject who cannot, who can never, fully recuperate the conditions of its own emergence.

But his death, if it is a death, is only the death of a certain kind of subject, one that was never possible to begin with, the death of a fantasy of impossible mastery, and so a loss of what one never had. In other words, it is a necessary grief.

This quotation recounts in a nutshell Butler's argument for an ethical responsibility based on 'an apprehension of epistemic limits' (Butler 2005: 43) concerning knowledge of the self. Instead of basing the ethical response on our capacity to provide a full account of oneself, Butler suggests that we must find a way to accept the fact that as we are not entirely transparent to ourselves we cannot provide the definite reasons for our actions.[10] This acceptance grounds a shared way of being in this world, and demands of us that we suspend our judgement of the other in the face of the opacity that constitutes not only my self but her self too. It entails that we recognise the extent to which we are *not* the authors of our actions, which are ultimately constituted in the unconscious and inaccessible limbo of the self, as it 'emerges' from the complex entanglements of linguistic and technological norms that constrain and make possible that we 'come about'.[11]

Butler briefly refers to psychoanalysis when asking us to attent to the opacity of an unconscious we cannot entirely retrieve for conscious introspection. Her discourse on what she calls 'the prehistory of the subject' (Butler 2005: 79) is an extended dialogue with Adorno, Nietzsche, Foucault, Hegel, Cavarero, Levinas, Kafka, Freud, Bollas, Klein, Laplanche, rather than an analysis of the findings and an exploration of the implications of what has been coined as 'The New Unconscious' within cognitive science.[12] Her dialogical monologue on the vulner-abilities of an exposed subject that stumbles upon its indeterminate, unfinished nature seems to turn Sartre upside down: the end of our life will not undo the enigma of who we are, finally or definitively, because the beginning that precedes us and constitutes us cannot be uncovered. Self-identity and a coherent account of who we are 'will have to fail in order to approach being true' (idem: 42),[13] 'the "I" cannot give a final or adequate account of itself because it cannot return to the scene of address by which it is inaugurated and it cannot narrate all of the rhetorical dimensions of the structure of address in which the account itself takes place' (idem: 67). However different the style of writing may be of cognitive scientists who wish to speak a clear and distinct language in which transparency reigns and philosophers who wish to speak a more imaginative language that discloses the ambiguity that springs from the opacity of the self, I think they are both – in their different idiom – referring to the same abyss from which our conscious experience wells up time and again. Butler indeed acknowledges that the opacity of our beginnings concerns *a reiterant event* ('prehistory does not stop happening', idem: 78), as we are continuously 'impinged' by others and by 'the way social force takes up residence within us, making it impossible to define ourselves in terms of free will' (idem: 106). What Butler accomplishes in her account of the impossibility of a full account of oneself is that she – probably unintentionally – provides a link between the detailed experiments of cognitive science that refute our sense of being sovereigns of our own self and philosophers and psychoanalysts who have been arguing for a relational conception of the self, while struggling with the implications for human agency, intentional action and individual responsibility.

If autonomous action was never a matter of undiluted control of a person over her own actions, how can we steer free of overconfident accounts of scientists claiming that free will is nothing more than an illusion? Must we agree with Nietzsche in lamenting the human condition, as he wrote more than a hundred years ago (1887) that people:

> felt unable to cope with the simplest undertakings; in this new world they no longer possessed their former guides, their regulating, unconscious, and infallible drives: they were reduced to thinking, inferring, reckoning, co-ordinating cause and effect, these unfortunate creatures; they were reduced to their 'consciousness', their weakest and most fallible organ![14]

Or, alternatively, is there good reason to celebrate the complexities of a conscious mind that is capable of self-reflection while being deeply – and to some extent safely – immersed in the inaccessible processes of autonomic behaviours that attune the organism to its internal and external environments? And could it be that to understand how this reflection came about and developed to its present form of reflexivity, we must turn to language and technology as constitutive for our capacity to externalise our thoughts, to create a distance from our self for the self and – in doing this – witness our birth as self-conscious beings (Ambrose 2001)?

If I tell a small child that she is Carol, whereas I am Mireille, by addressing her with: 'you are Carol' (pointing to her), and 'I am Mireille' (pointing to myself), she will initially answer me by confirming what I just said, pointing to herself and saying: 'you Carol', and pointing to me while saying: 'I Mireille'. I will retort, claiming 'no, *I* am Mireille and *you* are Carol'. At some point in time children learn to change perspective and suddenly realise that to others they are the 'you' that others are to them, while for themselves they are the 'I' that others claim to be themselves. This looking at the self from the perspective of the other is the birth

– in language – of the subject, the first person singular.[15] It is indeed *an affordance of language*. In that sense the subject is the consequence, though not in causal terms, of being addressed by the other(s), while this address is itself grounded in something that extends beyond the singularity of a particular address (language, social norms). To be born as an*other* and experience a first sense of self, however, is only the beginning of an adventure that has taken us from small-scale face-to-face communities to large-scale societies of strangers who are nevertheless constrained by shared social norms due to the jurisdictions to which they are subject(ed). We need to remember that the distantiation in space and time that conditions the appearance of large-scale societies has been an *affordance of the technology of the script*, while the introduction of the printing press allowed for even greater distantiation, thus giving birth to systemisation, indexation, rationalisation, visualisation and a concurrent turn to empirical science with its emphasis on visual representation (Eisenstein 2005, Goody and Watt 1963, Ihde 1990a, 1990b, 2002, Lévy 1990, Ong 1982, Ricoeur 1986). It seems that the script and the printing press afford an intensification of the externalisation and fixation of thoughts, making them available for reflection. This externalisation and fixation 'saves' them from the ephemeral quality of autonomic 'thoughts' as they appear in the conscious mind, thus reinforcing the process of translating autonomic thoughts into what I would call autonomous (reflective) thought.

If this is a fact, then the move to writing and printing must have triggered a shift to autonomy, to the possibility of giving an account of myself 'that not only explains what I do but allows me to assume greater agency in deciding what to do' (Butler 2005: 79). Indeed the move from speaking to writing and printing may be the 'cause' of our illusion of control, of willing our actions. Intentional action depends on conscious anticipation of the context in which we will act and of the consequences that our action may trigger. This anticipation is the fruit of deliberate reflection on past actions, allowing us to appropriate these actions and their consequences as *our* actions and as effects that *we* caused. Thus, in hindsight, we *become* the authors of these actions. However, in becoming the authors of our actions we also create the conditions for anticipating future actions and their consequences, thus creating the conditions for intentional action. This is not only related to the use of written and printed language, but already initiated by the interplay of remembering and anticipation made possible by language as such. In fact Ambrose (2001) describes, from an anthropological and archaeological perspective, how the use of tools presumes and produces some anticipation in terms of cause and effects, indicating that language may well be a necessary corollary of the instrumental usage of technologies that have turned us into animals that can take – and at some point in time can also articulate – an intentional stance in this world.[16]

This intentional stance is still rooted in the grounding opacity of the self, building on numerous autonomic processes and relying on inscribed (learned) pattern recognition and the related autonomic behavioural responses. Autonomy was not first, it cannot master its beginning, but this does not make it any less cause of celebration. It does not turn intentional action into an illusion; it only turns transparency and sovereignty of the self into an illusion. Autonomy is relative as well as relational. The self, as a body, as a mind that for the most is an array of subconscious autonomic processes to which its conscious dimension has no access, is exposed to permanent and pertinent impingement by the other(s), without ever having a final grip on its own development. But as this self arises from the depths of its subconscious and the width of its sociotechnical constitution, its relative and relational autonomy springs up, its natality (Arendt 1998) takes root and begins to influence both itself and its environment.[17] This is where, so far, autonomic computing is no match. Though the qualities of self-management, self-configuration, self-optimisation, self-healing and self-protection of autonomic computing require a certain spontaneity, randomness, unpredictability to be programmed into the system, this does not compare to a natality that guides self-conscious reflection. It may at some point approach the adaptive qualities of living organisms, but this does not equate to the conscious self-awareness that is preconditional for autonomous action. Autonomic computing may be vulnerable to destruction and to failure, but it has no reflective awareness of this vulnerability and it can for this reason not be held to account for the harm it causes.

Potential implications for criminal liability

Law and the ethics of judgement

The differences between ethics and law are manifold. What interests me here is that modern law requires more than Varela's ethical perception: it requires judgement. Varela's attempt to suspend the move towards judgement, which is structured in terms of the reasons that can be given for a particular action, is praiseworthy in as far as it concerns our capacity for *ethical* action. Butler's praise for an ethics of recognition that suspends judgement and takes into account the epistemic limits of our knowledge of the self and the other, is equally pertinent but seems similarly helpless when the *law* requires us to give an account of our actions in view of being charged with a violation of its norms. Modern law both assumes and produces an ethics of judgement. The judge must decide whether we have failed to live up to norms that exceed our particular situation, which were often already imposed upon us before we 'came about' and are consequently co-constitutive of our self.

Can we reject the charge with a story about the opacity or our self? Can we claim that we have no access to our autonomic mind and have no control over actions that may seem autonomous but are, in the end, 'caused' by inaccessible brain processes? Can the judge follow this logic and reply that her own autonomic programme directs her towards a conviction, because if we – who stand accused – are 'caused' by the unconscious, so is she? In the next subsection I will counter the argument that we were never capable of autonomous action anyway, implying that criminal liability in the end makes no sense. Equally pertinent for the subject of this volume, one could argue that the inevitable opacity of the unconscious dimensions of our mind dissolves the problem of autonomic computing environments having an invisible impact on our behaviour, since so much of our actions seem informed by invisible processes anyway. In the last section I will indicate how this relates to the next question, which concerns the idea that autonomic computing does not really make a difference, since we are already out of control.

We were never capable of autonomous action?

From our discussion of the unconscious brain processes that inform the interactions with our environment we have concluded that there is a fundamental opacity of the self that grounds us while we have no unmediated access to it. Nietzsche, Freud and Foucault have described the alien aspects of and within the self, long before cognitive psychology disclosed the untenable claims of the Cartesian ego. This ego is at least untenable at least in so far as it posits that when all distractions are peeled away we encounter a *res cogitans* that is capable of transparent rational thought about itself. From an ecological, relational perspective the distractions that hit us under the threshold of consciousness co-constitute our self and form an intractable layer of pattern recognition wired into the 'plastic' brain, perhaps reducing its plasticity while providing us with numerous sets of sophisticated well-attuned autonomic responses to similar events.

This, however, does not refute the possibility of autonomous action. Even if autonomous action depends on and produces autonomic action, it can be distinguished – though not separated – from the subconscious resources on which it nourishes. Although we cannot account for our own emergence as a human being in a particular time and place, the conscious self-awareness that is triggered by being addressed as the author of our action paves the way for intentional action. In that sense we can agree with Nietzsche when he claims that being called to account institutes self-reflection, delivering us to the task of providing reasons for our actions, even if we have little access to the correlated brain processes that are the precondition for any type of action.

Being called to account thus creates what it assumes. There is a problem here, which lies at the core of criminal liability. I would nevertheless argue that the circularity we encounter with regard to the institution of accountability is not vicious but virtuous, as long as we remember to incorporate the ethics of recognition that should precede an ethics of judgement. Keeping in mind that our roots are mostly invisible, untraceable and intractable we should be careful not to condemn a person beyond what she could and should have avoided. This care is not required because the accused was badly programmed but because we recognise a shared vulnerability to primary impingement by caretakers we have not chosen. This care does not imply that we stop calling to account those who misbehave, because we think their behaviour is caused by wrongly inscribed autonomic thought. On the contrary, this care implies that we acknowledge the importance of being called to account, as this is the most salient way to appropriate our actions as *our* actions and to generate autonomous action from the wells of autonomic thought.

The care that should inform legal judgement is inscribed in the procedural safeguards of the legal process. The fair trial entails the legally valid identification of a defendant as an offender and, if she is found guilty as charged, the trial enables the judge to call her to account. It provides for a delay, a hesitation, an extended period of time during which the charges are investigated in a court of law, instituting the possibility to contest the charges, claiming that one did not commit the incriminated act; or that the action one performed does not qualify as a criminal action; or that the action falls within the scope of the criminal law but can be justified or excused. The possibility to contest a charge takes the defendant serious as a legal subject, capable of acting in law, having 'standing' in court. Despite the temptation of treating the defendant as an object of investigation and punishment, the trial institutes legal constraints that force the judge to face the defendant as a person who, like the judge, is vulnerable to the inscriptions of her environment over which she had little control. The judge should suspend her judgement until she is lawfully convinced of the guilt of the defendant, otherwise – if this point is not achieved – the judge must acquit. And in facing the defendant the judge may develop an understanding of the other that clarifies what is opaque for this other, lost in the mist of early impingement by primary caretakers or by social norms that have imposed dangerous habits of the mind on the defendant. Habits that can explain her behaviour without necessarily excusing it. In convicting the offender the judge qualifies the other as an autonomous person, capable of resisting the imposition of social norms that invite behaviour that hurts or damages others to an extent that it violates the criminal law. But this is never an easy task, as it requires of a judge to be in touch with her own opacity, acknowledging that resisting autonomic behaviours can be extremely difficult.

The fact that our actions are grounded in autonomic behaviours over which we have limited control is not a reason to accept an appeal to being determined by autonomic brain processes. On the contrary, being called to account is preconditional for developing autonomous thought and for this reason the criminal trial is a pertinent occasion to generate autonomous action. In that sense criminal liability is productive of agency, to the extent that its attribution is organised in a fair trial that reminds the judge of the opacity she shares with the person she has to judge. In the meantime, the delay that is preconditional for an effective attribution of liability also relates to the capability of the judge to generate autonomous thought and to judge the other without being determined by her own autonomic thoughts. The findings of cognitive science confirm that the optimal decision is not the one taken 'on the spot' without any contemplation. However, these findings also suggest that the optimal decision may not be the one that is dictated by conscious deliberation, followed immediately by the decision. It would be interesting to develop experiments that test the excellence of more complex decisions that have normative implications and impact the life and well-being of others, to be taken after conscious deliberation, followed by an extended period of distraction, thus involving the unconscious brain in the process. Experiments that test the capacity of the unconscious brain to resist stereotyping (Glaser and Kilhstrom

2005: 186–89) indicate that autonomic thought is capable of applying normative constraints – somehow confirming Varela's proposition 'that a wise (or virtuous) person is one who knows what is good and spontaneously does it'. We must however qualify his statement by adding that in specific situations, such as the taking of decisions that may impact one or more others, spontaneity will only lead to 'good' actions if it is the result of conscious deliberation followed by a delay that allows for the unconscious mind to work things out, perhaps followed by a renewed deliberation, until a 'reflective equilibrium' is reached.

The final question: does autonomic computing make a difference?

Careful decision-making requires combining conscious deliberation with autonomic elaboration, trusting the unconscious mind to come up with relevant solutions, while taking the responsibility for the decision by returning to conscious deliberation. This return to conscious deliberation is triggered by the anticipation (the possibility) of being addressed as an autonomous agent, by being called to account. Though many decisions can and must be made in the flux of life, with no need and no time for such reflective interventions, autonomous action implies that at some point we must be open to debate on which course of action is either good or bad. We must, however, remember that we have no unmediated or complete access to the self, and need to be sensitive to the opacity of the roots of our thoughts. This opacity does not preclude us from opening our mind to what others may tell us about our self, providing us with a new view of our self and lifting the veil of our invisible motivations. Though we have no direct access to unconscious processes, the gaze of the other may inform us about our self, thus also co-constituting our sense of self, or self-consciousness. If this openness and the capacity to move back and forth between conscious and unconscious thought are the hallmark of autonomous action the question becomes relevant in which way autonomic computing systems make a difference to human identity and criminal liability.

Bateson once wrote that what matters is to detect 'the difference that makes a difference' (Bateson 1972: 315). Obviously many differences can be discerned between the human mind, building on her own autonomic thought and the human mind, building on processes of autonomic computing of an environment that proactively caters to her inferred preferences. In this text I have tried to describe the emergence of autonomous thought in the midst of our autonomic mind, thus preparing the ground for an investigation of 'the difference that makes a difference'. My guess would be that two differences may be relevant here. First, there is the matter of access to the way the autonomic environment profiles us: which opacity is exposed, correlated and disclosed by the profiling machines that mine our data? Second, the environment does not share this exposure with us. It does not share a fundamental opacity, because the environment is not (yet?) conscious of its own being. It does not expose itself, because it does not care to be exposed.

These two issues need further exploration, possibly providing new insights into our own vulnerabilities and opacity, as well as new ways to co-constitute the self.

Notes

1 See on the issue of criminal liability and Ambient Intelligence Hildebrandt (2008c).
2 This section draws on Hildebrandt 2008a, Hildebrandt and Gutwirth 2008, Hildebrandt 2009 and on exchanges within the workpackage on profiling of the EU funded Future of Identity in Information Society (FIDIS) Network of Excellence (NoE). See FIDIS deliverables 7.2/3/4/5/7/9/12/14 at www.fidis.net.
3 See the European Funded Coordinated Action (CA) on 'Cooperating Embedded Systems for Exploration and Control featuring Wireless Sensor Networks' at www.embedded-wisents.org.
4 See for instance: www.energy-retail.org.uk/smartmeters.html.
5 The usage of profiling techniques is not restricted to marketing and smart homes; it can also be adopted by forensic intelligence or elsewhere in the criminal justice system. See Harcourt 2007.
6 Representational in the sense that like certain ontological propositions are said to represent within the mind *the* reality as it 'is' outside the mind, certain deontological propositions can be thought to represent within the mind how actions should take place outside the mind (in the real world). This mentalistic picture is related to a correspondence theory of truth and depends on a strict separation of an inside and an outside, evoking the problematic of the homunculus. Profound criticism of this type of epistemology can be found, for instance, in Ryle (1949), Merleau-Ponty (1945), Varela et al. (1991), Gallagher and Zahavi (2008) and Clark (2003).
7 It is unclear whether Goldberg's use of the term 'analytic' refers to conscious analysis, or whether it also refers to information processing by the unconscious mind. I will use the term in reference to conscious deliberation and analysis, as will become clear below.
8 The idea of postanalytic autonomic thinking must not be taken to mean that we could ever consciously decide how to inscribe desired habits of thought and action into the neural pathways of the brain (Glaser and Kihlstrom 2005). For interesting descriptions of the impact of consciousness on habit formation see Bargh (2005: 51–54). See also Strack et al. (2006).
9 Let me be quick to note that previous or other societies must also be qualified as complex. However, with the advent of the script, the printing press and the digital era the scope of interhuman relationships has moved way beyond the face-to-face and the potential permutations and combinations of information, the ensuing de- and re-contextualisation as well as de- and re-territorialisation have increased the complexity of relationships between humans and nonhumans exponentially. See Lévy 1990, Kallinikos 2006.
10 This is an important clue to the importance of the right to privacy, which is what De Hert and Gutwirth (2006) have coined legal 'opacity tools'. The fact that we may want to shield knowledge about our selves from the gaze of others does not imply that we are transparent to our selves. Cf. Hudson (2006) who discusses privacy as a tool to protect *the secrets of the self.*
11 'Coming about' is also a sailing term which implies that the sailboat is turned by passing the bow of the boat through the wind. In this text I use it equivalently with 'emerging' and one could connect the two by remarking that the eruption of a self-consciousness is like taking a turn that is both continuous and discontinuous with non-reflective consciousness.

12   Her relational understanding of the self comes close to that of Mead (1959/1934), Ricoeur (1992) and Plessner (1975), all stressing the alterity that resides at the core of the self as well as the ex-centric constitution of the self.

13   This is why we must cherish the possibility to provide different accounts of our self, resisting efforts by others to impose a kind of self-identity that demands transparency and forbids what has been called 'the right to oblivion'. See saliently Rouvroy (2008: 252): 'Nous voulons soutenir que l'une des conditions nécessaires à l'épanouissement de l'autonomie individuelle est, pour l'individu, la possibilité d'envisager son existence non pas comme la confirmation ou la répétition de ses propres traces, mais comme la possibilité de changer de route, d'explorer des modes de vie et façons d'être nouveaux, en un mot, d'aller là où on ne l'attend pas.'

14   Quoted by Butler (2005: 11) from Friedrich Nietzsche, *On the Genealogy of Morals* (translated by Walter Kaufmann), New York: Random House 1969: 84.

15   't Hart (1995) has given a penetrating account of the fact that we are not born as subjects, referring to the work of the Swiss biologist Adolf Portmann (1887–1982), who emphasised that the biological birth of the child does not coincide with the birth of the person.

16   Intentional stance refers here to a capacity to have intentions, but in line with a relational understanding of human agency I would think that such a stance involves that we ascribe reasons to other actors. Cp. Dennett (2009).

17   For Arendt, human action entails an element of originality, spontaneity and freedom that cannot be reduced to an instrumental understanding of behaviours. This element, which she coined 'natality', seems to be on the verge of subconscious autonomic processes (which should not necessarily be understood as mechanical) and conscious appropriation.

References

Aarts, E. and Marzano, S. (2003) *The New Everyday. Views on Ambient Intelligence*, Rotterdam: 010.

Ambrose, S.H. (2001) 'Paeleolithic Technology and Human Evolution', (291) *Science*: 1748–53.

Arendt, H. (1998) *The Human Condition*, Chicago, IL: The University of Chicago Press.

Bargh, J.A. (2005) 'Bypassing the Will: Toward Demystifying the Nonconscious Control of Social Behavior', in Hassin, R.R., Uleman, J.S. and Bargh, J.A. (eds), *The New Unconsciousness,* Oxford University Press US: 37–58.

Bateson, G. (1972) *Steps to an Ecology of Mind*, New York: Ballantine.

Butler, J. (2005) *Giving an Account of Oneself*, New York: Fordham University Press.

Clark, A. (2003) *Natural-Born Cyborgs. Minds, Technologies, and the Future of Human Intelligence*, Oxford: Oxford University Press.

Custers, B. (2004) *The Power of Knowledge. Ethical, Legal, and Technological Aspects of Data Mining and Group Profiling in Epidemiology*, Nijmegen: Wolf Legal Publishers.

De Hert, P. and Gutwirth, S. (2006) 'Privacy, Data Protection and Law Enforcement. Opacity of the Individual and Transparency of Power', in Claes, E., Duff, A. and Gutwirth, S. (eds), *Privacy and the Criminal Law*, Antwerpen Oxford: Intersentia: 61–104.

Dennett, D. (2009) 'Intentional Systems Theory', in *Oxford Handbook of the Philosophy or Mind*, Oxford: Oxford University Press: 339–50.

Dijksterhuis, A. and Nordgren, L. F. (2006) 'A Theory of Unconscious Thought', (1) *Perspectives on Psychological Science* 2: 95–109.

Dijksterhuis, A. and Van Olden, Z. (2006) 'On the Benefits of Thinking Unconsciously: Unconscious Thought can Increase Post-choice Satisfaction', (42) *Journal of Experimental Social Psychology*: 627–31.

Eisenstein, E. (2005) *The Printing Revolution in Early Modern Europe*, Cambridge and New York: Cambridge University Press.

Fayyad, U. M., Piatetsky-Shapiro, G. et al. (eds) (1996) *Advances in Knowledge Discovery and Data Mining*, Meno Park, California – Cambridge, Massachusetts – London England: AAAI Press / MIT Press.

Franz, R., Simon, H. (2003) 'Artificial Intelligence as a Framework for Understanding Intuition', (24) *Journal of Economic Psychology*: 265–77.

Gallagher, D. and Zahavi, D. (2008) *The Phenomenological Mind. An Introduction to Philosophy of Mind and Cognitive Science*, London and New York: Routledge.

Gibson, J. (1986) *The Ecological Approach to Visual Perception*, New Jersey: Lawrence Erlbaum.

Gigerenzer, G. (2007) *Gut Feelings. The Intelligence of the Unconscious*, New York: Penquin.

Glaser, J. and Kihlstrom, J.F. (2005) 'Compensatory Automaticity: Unconscious Volition is Not an Oxymoron', in Hassin, R.R., Uleman, J.S. and Bargh, J.A. (eds), *The New Unconsciousness*, Oxford: Oxford University Press US: 171–95.

Goldberg, E. (2006) *The Wisdom Paradox. How Your Mind Can Grow Stronger As Your Brain Grows Older*, New York: Gotham.

—— (2006) 'The Wisdom Paradox', in Moody, H.R. (ed.), *Aging*, California: Thousands Oaks.

Goody, J. and Watt, I. (1963) 'The Consequences of Literacy' (5) *Comparative Studies in Society and History* 3: 304–45.

Haggard, P. and Libet, B. (2001) 'Conscious Intention and Brain Activity', (8) *Journal of Consciousness Studies* 11: 47–63.

Harcourt, B.E. (2007) *Against Prediction. Profiling, Policing and Punishing in an Actuarial Age*, Chicago and London: The University of Chicago Press.

't Hart, A.C. (1995) 'Mensenwerk? Over rechtsbegrip en mensbeeld in het strafrecht van de democratische rechtsstaat', (58) *KNAW Mededelingen van de Afdeling Letterkunde*, Nieuwe Reeks, 4 (Proceedings of the Dutch Royal Academy of Sciences).

Hildebrandt, M. (2008a) 'Profiles and Correlatable Humans', in Nico Stehr (ed.), *Who Owns Knowledge. Knowledge and the Law,* New Brunswick: Transaction Books: 265–85.

—— (2008b) 'Profiling and the Identity of the European Citizen', in Hildebrandt, M. and Gutwirth, S. (eds), *Profiling the European Citizens. Cross-Disciplinary Perspectives*, Dordrecht: Springer: 303–26.

—— (2008c) 'Ambient Intelligence, Criminal Liability and Democracy', (2) *Criminal Law and Philosophy* 2: 163–80.

—— (2009) 'Technology and the End of Law', in Claes, E., Devroe, W. and Keirsbilck, B. (eds), *Facing the Limits of the Law*, Dordrecht: Springer: 443–64.

Hildebrandt, M. and S. Gutwirth (eds.) (2008) *Profiling the European Citizens. Cross-Disciplinary Perspectives*, Springer: Dordrecht.

Hudson, B. (2006) 'Secrets of Self: Punishment and the Right to Privacy', in Claes, E. and Duff, A. (eds), *Privacy and the Criminal Law*, Antwerp Oxford: Intersentia.

Ihde, D. (1990a) *Technology and the Lifeworld*, Bloomington: Indiana University Press.

—— (1990b) *Instrumental Realism: The Interface between Philosophy of Technology and Philosophy of Science*, Bloomington: Indiana University Press.

—— (2002) *Bodies in Technology*, Minneapolis: University of Minnesota Press.

ISTAG (2001) 'Scenarios for Ambient Intelligence in 2010', Information Society Technology Advisory Group; available at ftp://ftp.cordis.europa.eu/pub/ist/docs/istagscenarios2010.pdf

ITU (2005) *The Internet of Things*, Geneva: International Telecommunications Union.

Jin, S., Park, S.-Y. and Lee, J.J. (2007) 'Driver Fatigue Detection Using a Genetic Algorithm', (11) *Artificial Life and Robotics* 1: 87–90.

Kallinikos, J. (2006) *The Consequences of Information. Institutional Implications of Technological Change*, Cheltenham, UK Northampton Edward Elgar.

—— (2008) 'Living in Ephemeria. The Shortening Life Spans of Information', available at: www.informationgrowth.info/docs/livinginephemeria.pdf.

Kephart, J.O. and Chess, D.M. (2003) 'The Vision of Autonomic Computing', *Computer*, January: 41–50.

Lévy, P. (1990) *Les technologies de l'intelligence. L'avenir à l'ère informatique*, Paris: La Découverte.

Loftus, E. and Ketcham, K. (1994) *The Myth of Repressed Memory*, New York: St Martin's Press.

Maturana, H.R. and Varela, F.J. (1991) *Autopoeises and Cognition: The Realization of the Living*, Dordrecht: Reidel.

Mead, G. H. (1959) *Mind, Self & Society. From the Standpoint of a Social Behaviorist*, Chicago: The University of Chicago Press.

Merleau-Ponty, M. (1945) *Phénoménologie de la Perception*, Paris: Gallimard.

Ong, W. J. (1982) *Orality and literacy: The technologizing of the word*, London: Methuen.

Plessner, H. (1975) *Die Stufen des Organischen under der Mensch. Einleitung in die philosophische in die philosophische Anthropologie*, Frankfurt: Suhrkamp.

Ricoeur, P. (1986) *Du texte a l'action. Essais d'herméneutique II*, Paris: Seuil.

—— (1992) *Oneself as Another*, Chicago: The University of Chicago Press.

Rouvroy, A. (2008) *Reinventer l'art d'oublier et de se faire oublier dan la societe de l'information? In La securite de l'individu numerise. Reflexions prospectives et internationales*, 249-78, Paris: L'Harmattan.

Ryle, G. (1949) *The Concept of Mind*, New York: Barnes and Noble.

Strack, F., Werth, L. and Deutsch, R. (2006) 'Reflective and Impulsive Determinants of Consumer Behavior', (16) *Journal of Consumer Psychology* 3: 205–16.

Varela, F.J., Thompson, E. and Rosch, E. (1991) *The Embodied Mind. Cognitive Science and Human Experience*, Cambridge, MA: MIT Press.

Varela, F.J. (1999) *Ethical Know-How. Action, Wisdom and Cognition*, Stanford: Stanford University Press.

Wegner, D. M. (2002) *The Illusion of Conscious Will*, Cambridge, MA: MIT Press.

Chapter 9

Technology and accountability
Autonomic computing and human agency

Jannis Kallinikos

Introduction

The ongoing infiltration of the social fabric by the technologies of computing and communication marks a distinctive stage in technology's involvement in human affairs. Technology and modernity are undeniably coextensive (Heller 1999). The patterns of living characteristic of the modern age have considerably been shaped by industrial technology and its far-reaching social and institutional impact. However, contemporary technologies of computing and communication differ in the sense of massively intervening upon the primary process of reality perception thus redefining the cognitive and communicative profile of daily living (Ayres 2007, Gantz et al. 2008, Kallinikos 2009b, 2009c). An important consequence is the constitution of human experience by the standardized and often unobtrusive procedures of assembling reality that current technologies of computing and communication mediate (Borgmann 1999, Manovich 2008). The immersion of technologies in the perceptual and decision space of daily life inevitably raises a range of issues as regards the degree of control and insight people have over the circumstances of their living. Much of what I say in this text is a response, a commentary and challenge to the claims Mireille Hildebrandt makes in her own chapter. In challenging her in an affirmative spirit, I hope to be able to qualify some of the ideas she pus forward concerning the implications which the current technological developments may have upon the ideals of *autonomous action* and the *accountability* of individuals that constitute essential pillars of the modern democratic and legal order.

The concerns Hildebrandt raises in her chapter are thus framed against the background of the negative prospects these developments may have for the self-determination of individuals and their ability to give an adequate account of the causes of their action. The information on which individuals draw in conducting their affairs is increasingly gathered or produced, processed and assembled into meaningful descriptions or patterns by means of methods, principles and procedures (automated data and information processing) on which ordinary people have little insight or control. Driven often by commercial interests, the versions of reality produced by these means often end up recommending courses of action that would have otherwise never been identified nor pursued by individuals. The complex entanglements of humans and technologies resulting from these developments thus call into question, Hildebrandt claims, the ideals of the sovereign individual and autonomous action and, in this process, undermine the construct of legal account-ability, an institutional pillar of modern law and democratic polity. Ultimately, what is at stake is the development of human pursuits on the basis of considerations that conform to the ideal of individual autonomy and express the free will of individuals. How is one to demarcate the space of individual sovereignty and distinguish individual autonomy and responsibility amidst the loops of descriptions, procedures and decisions in which humans and machines, individual pursuits and data processing algorithms fuse in ways that are difficult to disentangle? Can an individual still be called to give an account of her actions and choices on the basis of which her responsibility can be inferred?

The acknowledgement of the precarious nature of individual sovereignty and human autonomy is part and parcel of our cultural landscape and antedates the current technological developments. The coherent and omniscient self is a valorized description of the modern individual that descends from enlightenment and the worldview it has handed down to us. Over the last few decades, this idealized portrait of the self has persistently been questioned by several strands of philosophy and, more recently, by the findings of the sciences of cognition that reconstruct human perception and action as biologically driven. I therefore dedicate a section in which I trace Hildebrandt's critical reappraisal of the ideals of individual sov-ereignty and autonomy against the background of the questioning to which they have been subjected and I assess her effort to reclaim intentional action and individual accountability amidst the fragilities that afflict human life. In the last section of this chapter I take a step back from recent developments and reconsider some of the issues they engender by placing them in the wider historical context signified by the establishment of the modern social arrangement. The problematic of how to mould individual conduct, set free from the tight normative order of *gemeinschaft*, is key to the modern social arrangement (Heller 1999) and much can be learned, I contend, by placing current developments in this wider historical context. On the interpretation of modernity I offer, the social life and the self, society and the individual have been conceived and instituted as relatively inde-pendent spheres, and brought to bear upon one another along selective pathways that reinforce intentional action and individual and institutional accountability through the development of legible and simplified courses of action (Gellner 1983, 1995, Luhmann 1995, 1998). Rather than being a disembodied ideal, individuality is both a *structural principle* and an everyday *practice*. The characteristic modern solution of decoupling the self from the institutional order is, no doubt, endangered by the current infiltration of everyday living and the consequent transgression of the boundaries separating the self from the institutional order, which the tech-nologies of computing and communication enable.[1] However, at the same time as they undermine the modern solution, these same technologies and the practices to which they give rise provide considerable support to the project of constructing a legible world along which intentions can be plotted and followed up, essential to accountability and legal liability. The trade-offs underlying these trends are hard to predict but it is essential to recognize their double-edged nature. Prior to these undertakings, however, I give a brief outline of the developments associated with the diffusion and deepening social involvement of the contemporary technologies of computing and communication. While in broad agreement with Hildebrandt, the brief outline I attempt of recent technological trends allows the slightly different framing of the key issues in question.

Computation and human experience

After nearly half a century of computational developments, the profile of the change coinciding with the growing involvement of the contemporary technologies of computing and communication in social life remains blurred, shifting and contested. There are, certainly, plenty of reasons for viewing these developments in affirmative spirit (Anderson 2008, Ayres 2007). Computing machines, a wide range of handheld devices now in wide use and the communications networks through which they operate enable ordinary people to expand their experience creating or modifying, accessing and sharing a large variety of intellectual and cultural products or services such as news, ordinary or scientific texts, music and entertainment, photos, films or videos. At the same time, the multimodal virtual space that coincides with the spectacular growth of the internet, the convergence of media and the integration of global communications infrastructures holds the promise of breaking forever through the local confines of human experience.[2] Communities of people distributed over the globe emerge, prosper and dissolve tying together individuals in ways that celebrate individual proclivities and freedoms, shifting involvements or passionate concerns.

But there are plenty of justifiable reasons to be worried as well. Confines limit but also protect. A tedium as stability may occasionally be, it combats nonetheless the fugitive trends of human existence and anchors imagination and experience into the world. 'Routine', writes Sennett (1998: 43) cogently, 'can decompose labor, but it can compose a life'. These same developments and the promises they carry increasingly cast human living amidst a complex and, crucially, continuously shifting network of technical and social relations that may elude understanding and escape individual or institutional control. What is the status of the information that populates the new media and the internet? Is it meaningful, does it have a lasting value and can it be trusted? How are the constantly updatable versions of reality circulating in these virtual spaces produced, assembled, filtered and disseminated? Who is able to speak and who is not? How are freedom and privacy to be reframed and understood in a world in which human experience is suffused by information of all kinds and the protective belt of individual sovereignty loosens (Benkler 2006, Borgmann 1999, Hildebrandt and Gutwirth 2008)?

Some of these concerns are, of course, as old as the age around which the modern social order got consolidated. Modernity first detached life from the local bounds of *gemeinschaft* and cast it in the extended social aggregates of the newly formed nation states, and the anonymous, sizeable and permanently agitating urban spaces of modern cities (Gellner 1983, Giddens 1990, Heller 1999). Over roughly the same period, writing, printing and what we now call traditional media got increasingly integrated into private and public life (Eisenstein 1979). News and communications came increasingly under the control of the consolidated industry of mass media and powerful actors, including the state. As much as they are guaranteed by the institutional arrangements of modern polity, individual freedoms have in the course of modernity constantly been punctuated by the power of mass media. Over the last fifty years or so, in particular, the fugitive and spectacle-like reality of mass media has stepped into the forefront of social life, increasingly shaping the consciousness people develop of the world and of themselves (Baudrillard 1988, Bauman 2000, Luhmann 2000, McLuhan 1964).

Contemporary technologies of computing and communication introduce though important discontinuities that gestate novel opportunities and risks. For the first time in history, technology becomes directly and massively involved in the reconstruction and monitoring of processes that had until then remained the prerogative of the human mind.[3] From simple calculations to elaborate transaction recording systems, from databases to search engines, computer-based technologies enact automated rules, procedures or models (computer programmes) through which sign tokens (data and information) are acted upon, changed, stored and communicated. Prior to computation, these operations were certainly supported by a range of materials and tools but were exclusively performed by professions of various kinds. In modern administrative culture, in particular, the tasks of recording, storing and transmitting information were carried out by swarms of specialized clerical staff (Beniger 1986, Dreyfus 2001). While in cases such as photography, the recording of images entails the mechanical reconstruction (optics) of the vision process and its black-boxing, automation in traditional media remains low and crucially separated by technological islands (Kittler 1985). Whatever operations are accomplished by means of the technologies of traditional media (writing, filming or sound recording), they require the crucial presence of human agents and the operative skills these have interiorized and developed through long processes of learning, education and apprenticeship (Dreyfus 2001).

Little wonder, the operation of digital machines and the use of software require its own regime of rules and responsibilities. However, it is essential to understand that the massive involvement of computing in human affairs brings about a comprehensive reapportionment of the tasks and operations performed by humans versus machines. Increasingly, operations that would have previously been carried out by human agents are black-boxed and entrusted to machines. These developments provide the underlying technological conditions of the problems Hildebrandt identifies with respect to individual sovereignty and responsibility. For, how is the line that separates human responsibility from the blind enactment of computational rules to be drawn? How far can we rely on machine-run systems of disembodied rules that lack the human attuning to reality provided by the ongoing character of perception and cognition and the judgments they allow? What can and what cannot or should not be accomplished by means of these technologies? Over the past half a century, these concerns have perhaps acquired the most dramatic manifestations in the criticism directed against the project of *Artificial Intelligence* (see, e.g., Bloomfield and Vurdubakis 2008, Dreyfus and Dreyfus 1986, Hayles 2005).

However, the juxtaposition of human and artificial intelligence may invoke the deceptive imagery of a human-machine combat, gauged often, as robotics wishes, in anthropomorphic terms (Hayles 2005). Recent developments, I would like to point out, suggest that the prospect of machines replacing humans in a variety of fields come to pass through a slightly different route than the one imagined just a couple of decades ago. Rather than simply outperforming humans in resolving complex problems of technical nature, contemporary technologies of computing and communication diffuse throughout the social fabric to take over the monitoring of the trivia of life. Software-based artefacts of every kind increasingly infiltrate the minute details of everyday living. Routine, daily activities are increasingly conducted and carried out with the use of a variety of computer-based artefacts via the World Wide Web, the internet or other communications networks. In this process, a surreptitious but crucial shift occurs. Life situations tend to be defined as *cognitive problems* of computational nature to be resolved by the complex and automated computations performed upon the affluence of data and information tokens modern technologies and the life styles they instigate make available (Ayres 2007). It is vital to point out that none of the sophisticated technologies of the day can function without access to massive data that flow through the arteries of the increasingly global computing and communications infrastructures (Kallinikos 2006, 2009a). The pervasiveness of data and information is a resource and at the same time the problem to be solved. Search engines and web portals are clear manifestations of these trends.

By this route, the technologies of computing and communication break into the constitution of human experience in ways that have never been possible before. The spirit of these trends has been eloquently articulated, as this volume demonstrates, in IBM's speculative programme of *autonomic computing*. Half fiction, half reality, this programme envisages the convergence of several techno-cultural developments establishing a human habitat in which a substantial part of human experience is mediated by machines that scan the environment (including the human body) to track shifts (from room temperature to stock markets) that elude human perception, attention span, sensibility or memory and recommend (or initiate) accordingly suitable courses of action. Autonomic computing is still in its infancy and the market it seems to target concerns daily living only indirectly. However, a great deal of the technologies and practices it invokes have been around for some time now. Data mining and profiling, search engines, automatic identification technologies, semantic technologies are all in wide use.[4]

Obviously, the infiltration of the human habitat by the algorithmic reasoning provided by a seamless technological infrastructure connecting technologies, facilities and artefacts raises issues of transparency, control and privacy. It is perfectly reasonable to wonder about the implications these developments may have for autonomous action and the sovereign individual. Under the conditions described above, the versions of the world confronting people are assembled by a long series of computations about which they often have little insight and control. As noted in the introduction, this data-driven assembly of world versions is embedded in strong commercial interests and has serious implications for people's choices. Profiling technologies, in particular, assemble out of diverse data sources profiles of people (identities, lifestyles) and situations that are channelled back to people as incentives or motivations for acting in one way or another (e.g. buying a product or service or choosing a sex partner). In other words, classifications, patterns or profiles (especially commercially ones) that are produced by these means turn usually out interactive, in the sense Hacking (1999) attributes to the notion; that is, they are taken up by individuals, often unwittingly, and lived out as actions of their own choosing (Kallinikos 2009b).

But how novel are these developments? Have not human agents always acted on scant evidence, been agitated by seductive advertisements, incited to act in one or another way by information that could scarcely be called objective or impartial? Or, to go a little bit further, are not purpose, consistency and continuity posterior constructions by which human agents attribute meaning to the bewildering contingencies and events that populate their lives? As noted, contemporary philosophy and cognitive science construct a portrait of human agents driven by wider cultural forces, subconscious processes, semiotic and linguistic conventions over which they have little control or awareness. What new conditions do therefore the developments associated with the contemporary technologies of computing and communication bring about that further aggravate the precarious intentionality of human agents? I turn to the consideration of Hildebrandt's account of the challenge that current developments represent for autonomous and accountable action and her attempt to construct a weak version of the sovereign individual that rescues intentionality at the same time as it recognizes the frail constitution of human agency.

The self: coherence versus fragmentation

Legal accountability and the practice of calling someone to give an account of oneself assume individuals to be in command of the circumstances surrounding their choices. For, the ability to give an account of oneself presupposes access to the true reasons as well as knowledge and a degree of control over the causes, conditions and consequences of one's own actions. Over the last few decades, however, the view of the self as a coherent or conscious and in command of the circumstances facing it has been questioned by several branches of philosophy and more recently by cognitive science and the discovery of the subconscious, often hard-wired, processes governing perception, cognition and behaviour.[5]

As a matter of fact, the technological project of autonomic computing draws inspiration from cognitive science and the way the latter depicts the processes by which organisms map on their environment and adapt accordingly. The key concept in this regard is *pattern recognition*, that is, the mapping of environmental cues to a cognitive template that triggers a certain type or types of behaviour. Pattern recognition provides thus the overarching mechanism through which organisms sample their environment, infer or identify cause-effect relationships, pathways and other configurations as triggers to their adaptive behaviour. Pattern recognition occurs spontaneously and it is successful to the degree that entails a cognitive economy through which cues and situations are rapidly mapped onto the appropriate cognitive template (Bateson 1972, 1985, Goldberg 2006). These processes are preconceptual and subconscious in the non-psychological sense of invoking reactions that are often wired in the biological makeup of organisms. For this reason, they may be regarded as autonomic. In the human realm, pattern recognition reflects a melange of cultural and biological processes, even though the construct of culture is foreign currency to contemporary cognitive science. Drawing on Goldberg, Hildebrandt describes the development of perceiving habits and cognitive processes underlying pattern recognition as *postanalytic* to be distinguished from pre- or non-analytic behaviour. Postanalytic in this context means that the development of pattern recognition entails elements of reflection and evaluation of an intuitive rather than calculative nature through which the pattern is identified and gradually inscribed to the cognitive and behavioural repertory of individuals.

Thus described, pattern recognition is opposed to action through conscious deliberation associated with the ideal of autonomous subject and the liabilities such an ideal supports. But the scientistic language of cognitive science does not question the unity of the acting subject, at least not *prima facie*. For unity, consistency and coherence are cultural constructs. It is the human sciences and the arts that have accordingly painted the portrait of a fragmented and scarcely consequential self. History, language and culture constitute individuals as the particular persons they are at the same time as they transcend them in variety of ways (Heller 1999). While the outcome of social relations, very little of these historical and cultural processes and forces can be understood, shaped and controlled by the single individual. In any case, before one can interpret and understand reality as instantiated in a particular historical and cultural context, one is inexorably the product of that context. The recursive status of this relationship deprives humans as historically situated agents the stable and external reference from which they can view, understand and control their lives. To invoke such an Archimedean point is as paradoxical as the image of man climbing on his own shoulders (Valery 1956).

A key post-structuralist tenet, the elusive constitution of the self has perhaps been developed more eloquently in the works of Jacques Derrida and his persistent retracing (deconstruction) of meaning to a fugitive alterity that haunts whatever is presented as complete or solid, self-sufficient and clear (e.g. Derrida 1976, 1978, 1981).[6] On this view, whatever we write, say or do is gaining its identity or plenitude thanks to a complex, dynamic and constantly escaping fabric of differences. My utterance on a subject, for instance, that recommends itself as clear and solid, is made possible by what I (mostly unaware) exclude, what I do not say or know, the things that others have said or refused to say on this same subject, on the contrast of each block of meanings with other meanings, the distinction of each word I use with all other words, the play of differences underlying the constitution of syllables, sounds, characters, etc. These fugitive processes are obviously articulated differently in the various settings of social life and there may be social areas that have managed to negotiate a relatively stable reproduction of meanings and the practices to which they are associated. But in general, there is no way to trace this complex fabric of differences to a stable and external reference (a foundation) and to overdetermine its outcome. Widely known as it may be today, the deconstruction and decentring of human agency carries a counter-intuitive flavour that has definitely been the pole of attraction to many and dismay to the most.[7] The implications for the issues that concern us here should be rather straightforward. Language, culture, history, the unconscious are such inexhaustible fields of pulsating differences that traverse individuals from all directions, constitute them at the same time as they transcend and elude them in a variety of ways.[8]

Hildebrandt herself draws heavily on Judith Butler's relatively recent and evocative book *Giving an Account of Oneself*. Butler's (2005) book traces the cognitive, social and institutional conditions under which a reasonable account of oneself can be given. Its distinctive contribution notwithstanding, the argument concerning individual sovereignty and accountability Butler puts forward carries largely similar undertones to the ones described above. Only a precarious being in a complex, extensive and shifting network of relations provided by culture, language and the unconscious, individuals have no other than a fleeting and partial control over the circumstances of their lives. However, the limits to individual sovereignty and coherence, Butler claims, do not need to rule out the possibility of assuming ethical responsibility.[9] The fact that humans are ontologically constituted as individuals by wider social and cultural forces does not obviate the need of relating to these forces in ways that are singular and personal. The processes through which such a relationship is worked out are hermeneutic rather than causative (deterministic). Indeed, individuality and personhood to a large extent coincide with the different ways individuals address the circumstances and cultural forces by which they have originally been constituted. Drawing upon Agnes Heller, I will subsume this position under the notion of a reflective postmodernity (Heller 1999), that is, a modernity in which human living has lost forever confidence on the ideals of reason and progress but still has to rehabilitate a version of these ideals and live with the flaws and fragilities they imply.[10]

Even so, the implications of these claims for the ideal of sovereign and accountable individual are discouraging. If human behaviour is caught in a complex net of biological, linguistic, cultural and psychological forces that evade the awareness and conscious control of human agents, as these literatures suggest, then the practice of calling someone to give an account of his/her actions (legal/criminal liability) may lose much of its usefulness and suitability. Under these conditions, the difference between the versions of reality constructed by profiling and autonomic computing sketched in the preceding section and those triggered by subconscious psychological mechanisms, linguistic filters and deeply ingrained cultural habits is not clear. By extension, the implications for the construct of autonomous, intention-driven action assumed to be undermined by autonomic computing and the automation of daily living ought not to be qualitatively different from those derived from the deconstruction of the sovereign individual by philosophy and the nearly autonomic processes of perception, cognition and behaviour that cognitive science documents. Placed in this context, the problem Hildebrandt confronts is to make the difference between, on the one hand, the blind logic of algorithms of ambient computing that forges profiles of people through statistical permutations of data tokens and, on the other hand, the fragmented and unaware self that the deconstruction of the sovereign subject and the sciences of cognition have handed down to us. The problem is far from trivial. For, if there is no substantial difference between the two, then autonomic computing and profiling simply rewrap older problems and they accordingly cannot be blamed for attenuating the constitutional principles of modern society and polity. Let me elaborate.

While consigning transparency and control over the self to an illusion, the limitations afflicting individuals that philosophy and cognitive science identify may not rule out the possibility of intentional action. As Butler suggests, intentional action does not need to presuppose the illusion of the omniscient and omnipotent individual. All is needed is a reflection, no matter how precarious and incomplete, over one's own condition and the concomitant pursuit of objectives associated with the partial insight over the circumstances one confronts. Such an insight over one's own condition is produced out of the individual confrontation with the social, cultural and psychological forces that constitute each one of us and *inter alia* afforded by the objectification and externalization of the processes of cognition which language and writing/printing enable.

The capturing and stabilization of contingencies language affords, the storage, comparison and retrieval of recorded events writing and printing mediate may be used to obtain a modest degree of insight and eventually control over the circumstances of personal and social life (Goody 1977, 1986, Ong 1982). In addition to the review and systematization of the world which language and writing mediate, the objectification of cognition and thought they afford construct a bridge to the world that allows individuals to cross the distance to the social world, shift perspective and regard themselves from, as it were, the outside. To refer to oneself as 'me' implies seeing oneself from the standpoint of the other. The view therefore of the self (I) as an object (me) is already afforded by language. Such a shift in perspective may sharply raise the awareness one may obtain over oneself and the conditions one confronts (Mead 1934). In this regard, language and social life, Hildebrandt claims, ineluctably induce the steady crossing of one's own perspective by the perspective of the other, a process that ultimately reinforces the image the self acquires over itself. For, the self recognizes itself (identity, I) in the mirror of the other (difference, me). The view of the self from the standpoint of the other thus supports rather than undermines ego formation and clarifies some of the circumstances of living (Ricoeur 1992). In this sense, *memory* and *sociality*, *language* and *objectification* grant the affordances through which intention is crafted upon a fluid, incompletely known and fragmented self.

Intentional action and self-awareness are furthermore aided by the normativity of the very act of being called upon to give an account.[11] In this respect, the principle of legal liability brings about what *prima facie* seems to be predicated upon. Social expectations shape behaviour. The anticipation of the obligation to give an account in the case of law violation induces individuals to contemplate the reasons and consequences of their own actions. Furthermore, intentionality and self-awareness are positively related to the complex interrogative climate of the trial in which allegations can be contested and charges reframed and variously explored before they are finally applied. The trial is a context in which the contingencies and details underlying law transgression are meticulously examined, reasons provided and actions and decisions justified.

In this sense, autonomy (free will), says Hildebrandt, survives the fragilities that afflict individuals and the epistemic limits underlying their pursuits. It just is *relative* and *relational*. It is this weak yet essential version of autonomy that she feels is or may be undermined by the development of autonomic computing and computer-based profiling. She provides two groups of reasons for supporting her claims. 'First, there is the matter of access to the way the autonomic environment profiles us: which opacity is exposed, correlated and disclosed by the profiling machines that mine our data? Second, the environment does not share this exposure with us. It does not share a fundamental opacity, because the environment is not (yet?) conscious of its own being. It does not expose itself, because it does not care to be exposed'.

I applaud Hildebrandt's skilful navigation amidst the tidy waters of philosophy and cognitive science. The weak version of intentional action she assembles finds me in agreement. I think she thoughtfully summarizes our predicament in this historical juncture in which we have lost the certainties of modernity and become postmodern: that is, aware of the fact that we know as much as we ignore, that our motives are partly lucid and partly obscure and social complexity a nearly intractable condition which we must nonetheless accommodate. We still have to go on and carry out our projects amidst the uncertainties and ambiguities that rule our lives and we know are beyond repair (Heller 1999, Lyotard 1991, Vattimo 1989). I am also in broad agreement with the way she depicts the contemporary technological developments and the complex cognitive environment of data and information amassed by the technologies of computing and communication (Ayres 2007, Kallinikos 2006).[12] I am convinced by her claim that important principles and procedures of law born in the age of print must be revisited and presumably revised to accommodate the new realities of the information age. I find, however, the relationship between technology and individuality, objectification and intention in need of further elaboration. In the last section of this chapter I therefore outline some ideas that show that the technologies of computing and communication may under specific conditions provide a forceful ally to the project of constructing a

relatively legible world that may support rather than undermine individual autonomy and accountability.

Individual complexity and social life

In the weak version of autonomous action Hildebrandt constructs, language, speech
and writing emerge as key affordances through which the self articulates its relationship to the cultural, social and psychological forces that traverse and constitute
it and obtains awareness over its own actions and commitments.

This claim could perhaps be extended to the technologies of computing and
communication. While historically associated with writing and notation, the technologies of computing and communication carry into a new stage the techniques
and social practices for recording, indexing and storing information. A reflection
on schematic, *non-speech forms of writing* (such as the table, the list or the formula)
suggests strong affinities with the cognitive templates of computation and indicates
that standardized forms for recording reality have been widely used in history as
people sought to assemble the facts of life as pliable information (Goody 1977).
The implications of the deployment of writing methods and schemes for recording
reality seldom stops, however, at the instrumental purposes which such methods
and techniques serve. They feed back to social agents forming their cognitive and
perceptual habits and shaping their analytic skills. In this regard writing practices
constitute social agents. What we read and write affect who we are, how we think,
communicate and act. Much has been written on writing and its social implications
over the last half a century. McLuhan (1964) went indeed as far as to blame writing
for the linear reasoning and intellectual rigidity he attributed to western man. I will
not revisit that literature here (see, e.g. Bolter 1991, Eisenstein 1979, Ong 1982).
Positive or negative (and historical development entails many trade-offs), the
contribution of writing in forming the modern individual should be beyond any
doubt. It is indicative in this respect that Goody (1977) refers to writing and printing
as *technologies of the intellect* to indicate the significance they have had in the
historical construction of the kind of individual we are.

Similar to writing/printing, computer-generated information mediates states of
the world that provide the cognitive resources through which that world can be
contemplated, examined and compared, manipulated or acted upon. Information
produced and arranged by this means affords the distancing necessary to reflect
upon that world and the contingencies social agents confront, including their own
practices for dealing with these contingencies. There is little doubt that computing
differs from writing in a number of important ways manifested in the automated
(black-boxed) character by which data are processed and assembled to meaningful
units. Operations of this sort occur at a remove from people's awareness and
inspective capacity, a condition that is further aggravated by the comprehensive
nature of computerized data repositories. Recent advances in computation have
furthermore undermined the linearity of writing through the massive reintroduction
in daily living of images and the different semiotic logic of visual cognition and

communication.[13] These differences notwithstanding, it is reasonable to conjecture that the growing involvement of computing in social and personal life may promote, as Goody suggests writing has done earlier in history, the cognitive simplification and legibility of one's own life and, by extension, the pursuit of objectives under conditions that can be recalled, inspected, controlled or monitored. In all these respects, computing becomes a pervasive means through which individuals may articulate their relationship to the wider social and cultural environment in which they are embedded and obtain awareness over themselves and the conditions confronting them. Rather than being an adversary to the sovereign individual, the deployment of intellectual technologies in everyday life may accordingly contribute to constructing the legible cognitive space upon which intentionality can be mapped and carried out. This is how Luhmann (1998: 6–7) puts the issue:

> Technology, in its broader sense, is functional simplification, that is, a form of the reduction of complexity that can be constructed and realized even though the world and the society where this takes place is unknown. It is self-assessing. The emancipation of individuals, even irrational individuals, is an unavoidable side effect of this technologizing. . .this changes the social imperatives for individuality. The question is no longer 'What I should be?' but rather 'How should I be?' Whenever the individual is marginalized by technology in this way, a sense of distance is achieved that allows the individual to observe its own observation. . .An individual in the modern sense is someone who can observe his or her own observing.

An account of technology in these terms runs counter to the widespread belief (and attitude) that sees technology as basically an adversary to human autonomy and freedom. Such a belief descends from romanticism and has deep roots in humanities and the social sciences. Let me make clear that I stand myself dubious vis-à-vis technology in general and the current trends that are associated with the involvement of the technologies of computing and communication in everyday life (see e.g. Kallinikos 1996, 2006, 2009a). I have therefore no intention of putting up an advocacy for technology. I am aware of the fact that social change is always double edged and the developments that are associated with the growing infiltration of daily living by the technologies of computing are no exception. On the other hand, I feel that a defensive attitude vis-à-vis technology may fail to do justice to the central role it plays in constituting contemporary life forms. Technology is inextricably bound up with modernity, providing one of the constitutive logics of the modern social order (Heller 1999).[14] Such a fundamental condition needs to be contemplated and analyzed in terms other than technophobic.

Placed against this backdrop, the fear and anxieties that the discourse on privacy and control societies mediates (and no matter how loosely, Hildebrandt's text is associated with this discourse) are predicated on a view of individuality that may not be precisely compatible with the claims that accord technology a role constitutive of individuality. Ultimately framed in legal terms, the fear of control stems from a view of individual life that considers privacy and self-determination as the core of individuality. Privacy and self-determination are tied to a sphere of interiority, which the individual may want to give expression of in her own terms and protect it from others. Fencing off the private sphere of the individual from the threat of intervention by any external force (big or small brother!) that deploys the capabilities of contemporary technologies of computing and communication seems accordingly the proper response for protecting individual rights.

Reasonable and socially relevant as it often is, such an attitude may fail to observe that the link between the individual and society is far more elusive than it may seem in the first place. Much of what has been said so far in this chapter suggests that rather than being a disembodied ideal, individuality is a project in which persons are given the option (freedom, autonomy) but also, and crucially, the *tools* by means of which they become able to explore the social space in the discovery and pursuit of their objectives. It is thus important to point out that such a possibility is doubly constituted by the *freedom* and *capacity* of exploring. Freedom without capacity is an empty word. By the same token, exploration calls for a pliable or navigable social space whereby intentions can be plotted along courses of action that are legible and thus possible to pursue and ultimately assess or evaluate. Legibility is essential to exploration. In this sense, contemporary technologies of computing and communication could be seen as providing the means but also the micro-techniques (Foucault 1977, 1988) for the construction of such a navigable social space and its exploration. Prior to becoming adversaries, society and the individual need to construct that space upon which any conflict of interest is possible to recognize, trade off or suppress.

Stretching these claims a bit further, I would like to venture to suggest that the construction of navigable social spaces represents an important means through which society provides the institutional arrangements for dealing with the unpredictability and limited reliability of the individual. In other words, navigability and legibility are means for fashioning predictable and accountable modes of conduct. Modern life, in particular, fashions a distinctive mode of being a person (an individual), in which the complexities of psychological interiority are drastically separated from social conduct and the kind of social self the latter implicates (Gellner 1983, 1995). Psychological interiority is by and large a modern accomplishment that coincides with the separation and the decoupling of one's own fate (life journey) from hereditary (class and rank) social relations and the strong normative order of community. In this regard, modern people, as Heller (1999) suggests, become self-made, that is, able to set up their life journey by choosing where and how to live, which is another way of saying that they are able to articulate a distinctive (reflexive) relationship to the very conditions that constitute them as individuals in the first place (Butler 2005). But thus set free from the constraints of the tight normative order of *gemeinschaft*, modern people must be given the means to carry out their life projects in ways that are broadly speaking manageable and possible to assess from the point of view of the individual and the society as well. The legible character of the social space becomes then the essential mechanism through which individual interiority finds its way among the drives and motives that afflict its fugitive nature.

The story can perhaps be told in terms that are more primal or fundamental and superimpose upon the social/individual divide the subject/object dialectics. Objectification, of which technological arrangements are an instance, has always provided a counterpoint to the solipsist and aberrant trends of instinctual drives, reverie and contemplation. The solidity of objects and technological patterns stabilizes perception and anchors emotions. In this regard, things and technologies become the ontological mirror in which subjects recognize themselves, and experience and construct their limits and possibilities. This way modern society keeps at bay the shifting and unreliable motivational structure of the drives of human behaviour and the fragility of a psychological interiority, a condition that Freud, following perhaps Nietzsche (Butler 2005), explored in the dialectics of the ego and the id. Modern social life could thus be seen as that kind of arrangement in which personal motives and predispositions become streamlined and simplified. Such an account of social life may appear counter-intuitive, as it turns upside down the traditional view in which society is complex and individuals simple or, in any case, simpler. The issue is one though that is better gauged in terms of ambiguity/ predictability rather than complexity. Modern society can be viewed as that social arrangement that constructs relatively legible and structured (solid) action trajectories in which the fragile nature of individuality can find essential support, enabling thus individuals to pursue their life projects. While modern social life offers perhaps a much wider range of individual options and possibilities than any other social formation we know, each one of them is attained in the domesticated forms of tempered conduct that Albert Hirschman (1977) once subsumed under the notion of mildly calculative action driven by the 'lucid' character of interests rather than the abyss of instincts and passions.

Persisting issues

In this chapter I have sought to provide a few claims that support the idea that technological forms of constructing and monitoring one's own cognitive and cultural environment may, at least in some respects, strengthen rather than undermine individuality and intentional action. The growing involvement of computing technology in social life and the monitoring of the details that mark the course of everyday living may provide social agents with an array of capacities, reasons and resources that enhance their ability to deal with the variety of circumstances confronting them. Thus viewed, technologies of computing and communication can, similarly to writing, contribute to constructing a legible social space in which social agents are able to trace a variety of motives and situations, project and pursue their objectives and recall and retroactively track their actions or commitments. These conditions are conducive to forms of human agency that are considerably autonomous and accountable.

This is of course a general claim that does not address the issue of profiling and autonomic computing being used by institutional actors, such as corporations and the state, in ways that may undermine individual sovereignty and autonomy. Profiling and autonomic computing represent particular cases of the wider problem of transparency (opacity) and impartiality of the methods by which the versions of reality mediated by the technologies of computing and communication are assembled and disseminated. The more data and information become available over the internet and other information sources or networks, the greater the reliance on automated forms of data processing and ordering at a remove from people. In this regard, the issues of transparency and impartiality are endemic to the social and cognitive environment which the technologies of computing and communication help construct. This problem is presumably aggravated by the looping effects, which the profiles and patterns worked out by profiling methods may have upon social agents. However, by contrast to skewed and centralized character of previous technologies of information processing and communication (writing and broadcasting), the technologies of computing and communication offer greater leeway for lateral or community-based connections (as the internet and social networking demonstrate) that can disclose and undermine the premises by which powerful actors may strive to profit from the opportunities these technologies provide. Also the multiple sources of available information, the possibility to compare and crosscheck alternative courses of action may enhance reflexivity and the opportunity to, however modestly, evaluate or resist profiling recommendations.

The growing infiltration of everyday life by the technologies of computing and communication and the prospect of daily living being taken care by versions of autonomic computing are undeniably transposing meditation on accountability and freedom to an emerging realm in which new problems, risks and opportunities lurk. The issues Hildebrandt raises are persisting and it is quite possible they will come to grow in momentum in the years to come. They will join hand with trends such as the affluence of information, the over-stimulation and overabundance of options, the proliferation and expansion of the black-boxed processes on which autonomic computing and similar developments would be predicated and the crossing of institutional boundaries that safeguard the differentiation of individual and social life. Crucial among these developments would be the accelerating pace of time and the shortening life span of events whose relevant duration once conferred life meaning, purpose and continuity (Bauman 2000, Castells 1996, Heller 1999). Little wonder that a new world is on the making with lurking prospects that may not be altogether appealing. I have myself expressed my worries as regards these trends in a variety of fora (see e.g. Kallinikos 2009a, 2009b, 2009c). In this chapter I have though pursued another analytic path and sought to demonstrate (by reflecting on the thoughtful text of Hildebrandt) the double-edge nature of the current developments and the open, undecidable nature of social developments. Risks and opportunities are inextricably bound up with one another. An adversary as it may be to autonomous action, the growing involvement of technologies of computing and communication in everyday living gestates as well opportunities for enhancing social legibility, self-reflexivity and the capacity and skill to evaluate alternative courses of action, all of which are conducive to autonomous and accountable modes of action.

Notes

1 Recall the discourse on privacy and control societies.
2 By media convergence I mean the technological reintegration of the separate cultural traditions associated with text, image, sound and the genres they have given rise to (Bolter and Gruzin 2002, Kittler 1985, 1997, Manovich 2001, 2008). Convergence proceeds by means of the functional compatibility and interoperability of the various technologies and the cultural artefacts they produce.
3 By the term 'human mind' I do not necessarily mean individual cognition but all those intelligence-based or communicative operations that sustain social life and culture.
4 See www.research.ibm.com/autonomic/overview, also autonomic computing in *wikipedia* and Gilat (2005).
5 There is of course the older and recurring moral-philosophical debate of voluntarism versus determinism. Accounts of human behaviour in terms of contemporary cognitive science can be said to rewrap some of the older deterministic claims with evidence of an altogether different nature. Poststructuralist philosophy, however, articulates a distinctive kind of argument that rejects ultimate foundations thus featuring the importance which chance, ambiguity, complexity and lack of control have in human living.
6 Other key figures include names such as Roland Barthes, Jean Baudrillard, Michel Foucault, Jean-Francois Lyotard in France and Fredric Jameson and Richard Rorty in the US. Little wonder, the roots of these ideas can be traced back to key modern intellectual figures such as Freud, Nietzsche, Adorno and Heidegger.
7 Derrida often refers to this project as critique of logocentrism and western metaphysics.
8 Some of these ideas have in fact been debated rather intensively in the less exotic interdisciplinary discourse on rationality (predominantly north American) that has questioned the omniscient and omnipotent qualities of human agents presupposed by the liberal ideals of utility maximization and goal accomplishment. See, e.g. Elster (1984), March (1994), Sen (1987). For a summary of that literature and its connection to continental thought see Kallinikos (1996, 1998).
9 None of the poststructuralists I know would deny the ethical responsibility that is associated with our choices. What they mostly debunk is the arrogance of the belief of knowing of being right and the idea of the world such a belief invokes.
10 Such a position stands indeed close to the idea of postmodernity as originally described by Lyotard (1984, 1991).
11 Butler (2005) traces this back to Nietzsche and his assertion put forth in *On the Genealogy of Morals* that reflection over the self which accountability presupposes results from accusation.
12 See also the special issue of *Wired* 'The End of Science', July 2008.
13 See Flusser (2000) for an interesting account of the cognitive and cultural processes signified by the technological return of the image.
14 Heller (1999) describes three constitutive logics of modernity, the logic *of technology/ science*, the logic of *social division of labour* and the logic of *political power*.

References

Anderson, C. (2008) 'The End of Theory', *Wired*, July 2008.

Ayres, I. (2007) *Super Crunchers: How Everything Can be Predicted*, London: Murray.

Bateson, G. (1972) *Steps to an Ecology of Mind*, New York: Ballantine.

—— (1985) *Mind and Nature*, London: Flamingo.

Baudrillard, J. (1988) *Selected Writings*, Stanford: Stanford University Press.

Bauman, Z. (2000) *Liquid Modernity*, Cambridge: Polity.

Beniger, J. (1986) *The Control Revolution: Technological and Economic Origins of the Information Society*, Cambridge, MA: The MIT Press.

Benkler, Y. (2006) *The Wealth of Networks*, New Haven: Yale University Press and www.benkler.org (creative commons license).

Bloomfield, B. and Vurdubakis, T. (2008) 'On IBM's Chess Player: On AI and its Supplements', (24) *The Information Society*, 2: 69–82.

Bolter, J. D. (1991) *The Writing Space: The Computer, Hypertext and the History of Writing*, London: LEA Publishers.

Bolter, J. D. and Gruzin, R. (2002) *Remediation: Understanding New Media*, Cambridge, MA: The MIT Press.

Borgmann, A. (1999) *Holding on to Reality: The Nature of Information at the Turn of the Millennium*, Chicago: The University of Chicago Press.

Butler, J. (2005) *Giving an Account of Oneself*, New York: Fordham University Press.

Castells, M. (1996) *The Rise of Network Society*, Oxford: Blackwell.

Derrida, J. (1976) *Of Grammatology*, Baltimore: The John Hopkins University Press.

—— (1978) *Writing and Difference*, London: Routledge.

—— (1981) *Positions*, Chicago: The University of Chicago Press.

Dreyfus, H. (2001) *On the Internet*, London: Routledge.

Dreyfus, H. and Dreyfus, S. (1986) *Mind over Machine*, New York: Free Press

Eisenstein, E. (1979) *The Printing Press as an Agent of Change: Communications and Cultural Transformations in Early Modern Europe*, Cambridge: Cambridge University Press.

Elster, J. (1984) *Ulysses and the Sirens: Studies in Rationality and Irrationality*, Cambridge: Cambridge University Press.

Flusser, V. (2000) *Towards a Philosophy of Photography*, London: Reaktion Books.

Foucault, M. (1977) *Discipline and Punish: The Birth of the Prison*, London: Penguin.

—— (1988) 'Technologies of the Self', in Martin, L. H., Gutman, G. and Hutton, P. H. (eds) *Technologies of the Self*, London: Tavistock.

Gantz, J. F. et al. (2008) 'The Diverse and Exploding Digital Universe: A Forecast of Worldwide Information Growth Through 2011', *IDC White Paper* www.emc.com/digital_universe, accessed 22 April 2010.

Gellner, E. (1983) *Nations and Nationalism*, Oxford: Blackwell.

—— (1995) *Conditions of Liberty: Civil Society and its Rivals*, London: Penguin.

Giddens, A. (1990) *The Consequences of Modernity*, Cambridge: Polity.

Gilat, D. (2005) 'Autonomic Computing-Building Self-Managing Computing Systems', in Flachbart, G. and Weibel, P. (eds) *Disappearing Architecture*, Basel: Birkhauser.

Goldberg, E. (2006) *The Wisdom Paradox. How Your Mind Can Grow Stronger as Your Brain Grows Older*, New York: Gotham.

Goody, J (1977) *The Domestication of the Savage Mind*, Cambridge: Cambridge University Press.

—— (1986) *The Logic of Writing and the Organization of Society*, Cambridge: Cambridge University Press.

Hacking, I. (1999) *The Social Construction of What?* Cambridge, MA: Harvard University Press.

Hayles, C. (2005) 'Computing the Human', *Theory, Culture and Society*, 22/1: 131–51.

Heller, A. (1999) *A Theory of Modernity*, Oxford: Blackwell.

Hildebrandt, M. and Gutwirth, S. (eds) (2008) *Profiling the European Citizen*, Dordrecht: Springer.

Hirschman, A. (1977) *The Passions and the Interests*, Princeton: Princeton University Press.

Kallinikos, J. (1996) *Technology and Society: Interdisciplinary Studies in Formal Organization*, Munich: Accedo.

—— (1998) 'Utilities, Toys and Make-Believe: Remarks on the Instrumental Experience', in Chia, R. (ed.) *In the Realm of Formal Organization: Essays for Robert Cooper*, London: Routledge.

—— (2006) *The Consequences of Information: Institutional Implications of Technological Change*, Cheltenham: Elgar.

—— (2009a) 'On the Computational Rendition of Reality: Artefacts and Human Agency', (16) *Organization*, 2: 183–202.

—— (2009b) 'The Making of Ephemeria: On the Shortening Life Spans of Information', *The International Journal of Interdisciplinary Social Sciences*, {4/3}: 227–36.

—— (2009c) 'Attempting the Impossible: Constructing Life out of Digital Records', www.telos-eu.com/en/article/attempting_the_impossible_constructing_life_out_ (8 July 2009), accessed 22 April 2009.

Kittler, F. A. (1985) *Gramophone, Film, Typewriter*, Stanford: Stanford University Press.

—— (1997) *Literature, Media, Information Systems*, Amsterdam: OPA.

Luhmann, N. (1995) *Social Systems*, Stanford: Stanford University Press.

—— (1998) *Observations on Modernity*, Stanford: Stanford University Press.

—— (2000) *The Reality of Mass Media*, Cambridge: Polity Press.

Lyotard, J.-F. (1984) *The Postmodern Condition: A Report on Knowledge*, Manchester: Manchester University Press.

—— (1991) *The Inhuman*, Cambridge: Polity.

Manovich, L. (2001) *The Language of New Media*, Cambridge, MA: The MIT Press.

—— (2008) *Software Takes Command*, http://lab.softwarestudies.com/2008/11/softbook.html, accessed 22 April 2010.

March, J. (1994) *A Premier on Decision Making*, New York: Free Press.

McLuhan, M. (1964) *Understanding Media: The Extensions of Man*, London: Routledge.

Mead, G. H. (1934) *Mind, Self and Society*, Chicago: The University of Chicago Press.

Ong, W. J. (1982) *Orality and Literacy: The Technologizing of the Word*, London: Routledge.

Ricoeur, P. (1992) *Oneself as Another*, Chicago: The University of Chicago Press.

Sen, A. (1987) *On Ethics and Economics*, Cambridge: Cambridge University Press.

Sennett, R. (1998) *The Corrosion of Character: The Personal Consequences of Work in the New Capitalism*, New York: Norton.

Valery, P. (1956) *Dialogues*, Princeton: Princeton University Press.

Vattimo, G. (1979) *The End of Modernity*, Cambridge: Polity.

Chapter 10

Of machines and men

The road to identity. Scenes for a discussion[1]

Stefano Rodotà

1. In the reflections on the 'homme-machine' by La Mettrie and D'Holbach (de La Mettrie 1987, D'Holbach 1990), the physical and psychical identity is ordered, in a regulatory sense, by nature. But it is the relationship with the world of machines that shows that identity is a complex social entity, irreducible solely to a naturalistic data, resulting from a never accomplished historical event. Montaigne reminds us that life, in which identity is reflected, 'est un mouvement inégal, irrégulier, multiforme',[2] thus a continuous construction, entrusted to variable contexts, departing from any automatism. Furthermore, if the order that governs identity were only naturalistic, then the autonomy of a person itself would be denied at its origin.

Rather, throughout history we have always tried to force the limits of nature, especially when we have tried to mimic it, reproduce it, transport it to a different dimension. It is not a paradoxical conclusion, but just when reproduction of nature appears at its zenith, the highest degree of artificiality has been reached. The automatons, the Ingenious Devices have fascinated us since ancient times; they have paved the way to other mechanical creatures, like robots and the different thinking machines; and then came the cyborgs, announcing the trans- and post-human, the researches on brain-machine interfaces (BMIs) or brain-computer interfaces (BCIs). But relations between man and the world of machines are not linear (Losano 1990, Sini 2009). The fact that we start from man as a reference or model may lead to very different results: to try to replicate man in a machine or to replicate machine in a man, an object among other objects, in fact an 'homme-machine'.

Social concern has always accompanied these matters; it has brought about many reactions against, and radical criticism of, mechanicism; the most well-known and extreme reaction was the one that went under the name of Luddism, which manifested itself still in the '60s when Harvey Matusow and his International Society for the Abolition of Data Processing Machines demonstrated in front of the IBM offices, raising banners saying: 'Computers Are Obscene'. Criticism of technological progress has expressed itself in various forms, to protect man from a fate that makes him become 'a happy slave of machines' and to prevent the whole of society from changing into a relentless control machine.

In reality, great dystopias and utopias constitute an essential cultural background at the basis of any discussion on the man-machine relationships. Turnarounds are continuous. It is the progressive integration with the world of science and machines, in fact, that has been considered also as an extraordinary chance given to the world of humans to reach a fullness that it had lacked until then. The *Magnalia naturae* listed by Francis Bacon at the end of *New Atlantis* comes back to the mind:

> the prolongation of life; the retardation of age; the curing of diseases counted incurable; the mitigation of pain; the altering of temperament, the statures, the physical characteristics; the increasing and exalting of the intellectual parts; versions of bodies into other bodies; making of new species; drawing of new foods out of substances not now in use.[3]

All this, today, can also be considered in the dimension of rights, a construction of identity that ends up coinciding with the construction itself of humans. After the Oskar Pistorius issue, the South African runner running with two carbon fibre artificial limbs replacing his lower legs (authorised to participate into the Olympic Games), another Paralympic athlete, Aimée Mullins, said that 'to change one's body through technology is not an advantage, but a right. Both for those doing sports professionally and common people' (2008). Thus the barrier between 'normal people' and those with artificial prosthesis falls, and in fact a wider notion of 'normal' is developed, which becomes a condition to freely construct one's identity using all the socially available opportunities. The new dimension of humans calls for a different legal measure, that can dilate the scope of the fundamental human rights.

Through the body, people can take possession of technology, bringing it back to the dimension of humans. But what happens when these phenomena do not manifest themselves as appropriation, but as expropriation, when people find themselves living in an environment where machines can take over their identity, change their body to enable its external control, when we live in an augmented reality, ambient intelligence, ubiquitous or pervasive computing, smart environments, when, all said and done, we live in an environment where machines can take on a position of supremacy, for whatever use is made of them or for their own autonomic nature?

2. To try and answer these questions, and grasp the new way in which identity is constructed, we have to start from the realisation that a social and legal order of machines is developing that claims its own autonomy, that not only may clash with people's traditional autonomy but also give rise to a new anthropology. Two judgements of the Bundesverfassungsgericht (the German Constitutional Court) may help clarify some aspects of the problem.

The first (15 February 2006) concerns par. 14.III of Luftsicherheitgesetz,[4] that authorised military aviation to shoot down a civilian plane as, after having been hijacked by terrorists, it was feared that it could be used as a weapon against civilian or military targets (the case of the 9/11 attack to the Twin Towers and the

Pentagon), without there being any other way of preventing such an outcome. The said rule was considered in contravention with Articles 1 and 2 of the Grundgesetz, on the dignity and defence of life, two particularly significant grounds. The German constitutional judges, in fact, thought that the passengers of the aeroplane were being 'depersonalised and, at the same time, deprived of their rights' ('verdinglicht und zugleich entrechtlicht') (Section 124). A unilateral decision by the State on the life of the passengers deprived them of the right of every human being to autonomously decide of its own existence. They were reduced to inanimate objects, to mere components of the plane, to be absorbed by the machine and undergoing a radical change in their prerogatives, in their 'human' status.

Just as important is the subsequent judgement of 27 February 2008,[5] by which the Bundesverfassungsgericht declared in contravention with the Grundgesetz an amendment to the law about the domestic intelligence service of the Land North-Rhine Westphalia. The amendment had introduced a right for the intelligence service to 'covertly observe and otherwise reconnoitre the Internet, especially the covert participation in its communication devices and the search for these, as well as the clandestine access to information-technological systems among others by technical means' (Section 3). The decision of the Bundesverfassunsgericht is widely considered a landmark ruling, because it constitutes a new 'basic right to the confidentiality and integrity of information-technological systems' as part of the general personality rights in the German constitution. The reasoning goes: 'From the relevance of the use of information-technological systems for the expression of personality (Persönlichkeitsentfaltung) and from the dangers for personality that are connected to this use follows a need for protection that is significant for basic rights. The individual is depending upon the state respecting the justifiable expectations for the integrity and confidentiality of such systems with a view to the unrestricted expression of personality' (margin number 181). The decision complements earlier landmark privacy rulings by the Constitutional Court that had introduced the 'right to informational self-determination' (1983) and the right to the 'absolute protection of the core area of the private conduct of life' (2004).[6]

Information-technical systems that are protected under the new basic right are all systems that 'alone or in their technical interconnectedness can contain personal data of the affected person in a scope and multiplicity such that access to the system makes it possible to get insight into relevant parts of the conduct of life of a person or even gather a meaningful picture of the personality' (margin number 203). This includes laptops, PDAs and mobile phones.

The decision also gives very strict exceptions for breaking this basic right. Only if there are 'factual indications for a concrete danger' in a specific case for the life, body and freedom of persons or for the foundations of the state or the existence of humans, government agencies may use these measures after approval by a judge. They do not, however, need a sufficient probability that the danger will materialise in the near future. Online searches can therefore not be used for normal criminal investigations or general intelligence work.

Let us compare this approach with the one that emerges from the front cover of the first issue of the 2007 Time's magazine, dedicated as is tradition to 'the person of the year', which spelled out in big letters the word 'You'. Thus it was the endless numbers of individuals to be indicated as the protagonist. Each one, though, in their unrepeatable individuality, because the front page was made of reflecting material that allowed anyone looking at it to recognise himself as in a mirror. You are the world.

But, at a closer look, that mirror was a computer screen drawn over the word 'You'. The message thus takes on a particular meaning. I recognise you as the person of the year because you have become part of the most significant technological apparatus. The man-machine order is upside down. You are a protagonist, and maybe the lord of the environment around you, only if you become a machine yourself, basically if you become a part of the apparatus.

In the German judgement of 2008 this approach is completely overturned. It is the human that embodies the machine, not the opposite. It is recognised that between man and machines not only is there an interaction, but a compenetration. This is a structurally evident data, and its constitutional relevance is recognised. The law thus reiterates the priority of humans, but manifests its power telling us that the world is going through a new entity, made up of the person and the technical apparatus to which data are entrusted. A continuum is established between the person and the machine: by recognising this, the law hands us a new anthropology, affecting legal classifications and changing their quality. Confidentiality, a quality of humans, is handed over to the machine. But this new fundamental right is now challenged by the cloud computing. Remoteness, separation, depersonalisation of this infrastructure produce a new form of separation between the man and the machine.

It is not possible, then, to consider this judgement as just a further development of the orientation adopted by the Constitutional Court itself in 1983. By that historic judgement, the Constitutional Court recognised 'informative self-determination' as a fundamental right, and radically changed the traditional picture of privacy protection, retrieving the more important indications of the cultural development started in the early '70s. In the 2008 judgement reference to confidentiality still appears, even if its transfer from the person to the machine already confirms a new approach. But two fundamental questions differentiate it from the previous ruling. The first concerns the fact that the notion of confidentiality is extended to comprise an ensemble that prevents reducing the person, and thus the human, to a mere material entity. From here, and this is the second important point, there is a new form of protection, that overcomes the dychotomy between *habeas corpus,* linked to the physical body, and *habeas data,* conceived as an extension of that historical protection of the electronic body. We no longer have separate entities to be protected but only one: the person in its different configurations, progressively determined by his relation with technologies, which are not only electronic.

We are facing the reconstruction of the integrality of the person, similar to the one realised through the recognition of a unitary protection of the person's integrity, no longer limited only to a physical integrity, but extended to comprise also psychical and social integrity, as explicitly set forth in the definition of health developed by the World Health Organization ('a state of complete physical, mental and social well-being and not merely absence of disease or infirmity') and then transposed in a multitude of legal documents (e.g. Article 3 of the Charter of Fundamental Rights of the European Union). We could say, with some rhetoric emphasis, that the law, after having acknowledged the non-severability of the body and the soul, provides its version of 'man machine' through the 2008 German judgement. Primary importance is given to the human element, the only way to reconcile man with the technical apparatuses that progressively accompany, restructure and invade him.

Personal identity is expanding. But who, concretely, is behind this? The answer emerging from the 2008 German judgement seems to tell us that it should always and only be the person concerned to set the conditions for defining his identity, by re-establishing his rule on a portion of the external world, and the technical apparatuses he directly uses.

That has never been the case; the construction of identity cannot be confused with the right to self–representation. But the technological changes in the methods for processing personal information have progressively changed the relationship between the identity freely constructed by the individual and the intervention of third parties, giving growing weight to the activities of the last mentioned.

Inaccuracies and partial representations, or even real falsifications, are a constant feature of many biographies freely developed by entities other than the person concerned, which then become part of the socially accredited information structures (like Wikipedia). Nowadays we also have 'dispersed' identity, in the sense that information concerning the same person is entered in different data banks, each one of which only returns a part or a fragment of the overall identity. We risk entering a time of identity 'unknown' to the person concerned himself, in the sense that not only is it found in different places, but also in places that are difficult or impossible to know of, or to have access to.

Our identity, thus, is more and more the result of an operation prevailingly conducted, processed and controlled by others. And we are not speaking here only of a construction based on the way the others see us and define us: 'le Juif dépend de l'opinion pour sa profession, ses droits et sa vie' (Sartre 1954: 106). Collective representation can determine the way in which we are considered, without necessarily providing the identity constitutive materials, as it happens when personal data are used directly. Furthermore, it is also true that in the one or the other case we have an 'unstable' identity, at the mercy, from time to time, of moods, prejudices or the concrete interest of the entities collecting, storing and disseminating personal data. A circumstance of dependence is thus created that causes the construction of an 'external' identity, and the classification of identity in forms that reduce the identity managing power of the person concerned.

3. A small American story can help us understand a change that has already become a part of our daily lives. In a primary school of California, for security

reasons, it was decided that each child would carry a necklace including a smart tag – a remotely readable RFID chip – to enable charting all his or her movements and localise him or her at any time. Back home, a little girl commented that novelty like this with her parents: 'I do not want to become a packet of cereals'. The reference made here is clear. It reflects the experience of a girl who, in a super-market, sees that the things she buys are 'read' by an electronic instrument through the bar codes, and refuses to be assimilated to a mere remotely readable object.

The little girl has understood everything, she has described a change that affects not only society, but the very anthropological nature of individuals. The body is growing in its importance, and is changing its functions. It is becoming a source of new information and is exploited by unrelenting data mining activities – it is a veritable open-air mine from which data can be extracted uninterruptedly. The body as such is becoming a *password* – physicality is replacing abstract passwords. Fingerprints, hand/finger/ear geometry, iris, retina, face scans, body odours, voice, signature, keystroke dynamics, gait recognition, DNA are increasingly used as biometric information not only to identify individuals or allow access to several services, but also to set up permanent categories and perform additional controls following identification or authentication/verification, that is, confirmation of someone's identity.

The body is getting into focus. A manipulated body is being created, which is primed towards surveillance and can be located. Technology is working directly on the body. Surveillance is no longer implemented from the outside, for instance by means of video surveillance. It is not enough to exploit physical features, as is the case with the use of biometric data. Indeed, the body is coupled with electronic devices – first and foremost, those based on the RFID technology. The body is supplemented and modified by means of the insertion of electronic implants and of the use of nanotechnologies. The body is being transformed as a whole, not only because it is becoming post-human or trans-human, but because the very autonomy of individuals is being affected – since individuals can be monitored and directed remotely. The body is therefore becoming a new object, which also requires considering anew what is a personal data nowadays in order to ensure that the protection of such data as currently envisaged remains workable, and personal identity can remain under control by the subject itself.

We have concrete examples before us, and they are growing in number by the day. We all are aware of the cases of employees required to carry a small 'wearable computer', which allows the employer to guide their activities via satellite, direct them to the goods to collect, specify the routes to be followed or the work to be done, monitor all their movements and thereby locate them at all times. In a report published in 2005 by Professor Michael Blackmore from Durham University, commissioned by the English GMB trade union, it was pointed out that this system already concerned thousand people and had transformed workplaces into 'battery farms' by setting the stage for 'prison surveillance'. We are facing a small-scale Panopticon, the harbinger of the possibility for these types of social surveillance to become increasingly widespread. Similar results, although concerning only location at the workplace, are already possible by means of the insertion of an RFID chip in employees' badges.

Some companies, like City Watcher in Ohio, took another step forward in the direction of manipulating its employees' bodies by requiring some of them to have a microchip implanted in their shoulders in order to be identified at the entrance of restricted access areas. Thus, the body is modified in its very physicality and primed in order to be monitored directly. And the body implants of remotely readable microchips are being increasingly used in the most diverse sectors, from disco clubs to hospitals, to open the door of one's house or start up one's computer – with a resulting cost reduction and growing ease of deployment.

In some countries such as Italy the application of these technologies is prohibited where it results in employees being monitored remotely. However, it is not enough to propose that this ban should become general and set out as a rule applying to the whole European Union. Indeed, these technologies are also used in respect of individuals and activities unrelated to the employment sector: therefore, it is necessary to tackle the issue of the lawfulness of using tools that entail manipulation of the body. In the opinion rendered in 2005 by the European Groups for Ethics in Science and New Technologies of the European Commission, concerning ITC implants in the human body, it was concluded that the use of microchips was admissible only in limited cases and exclusively in order to safeguard the data subjects' health. Any other utilisation should be regarded as in conflict with human dignity, which was declared inviolable in Article 1 of the Charter of fundamental rights of the EU, as well as with data protection principles.

What would be of a society in which a growing number of individuals were tagged and tracked? Social surveillance is committed to a sort of electronic leash. The human body is equated to any moving object, which can be monitored remotely via satellite technologies or else radiofrequency devices. If the body can become a password, location technologies are bringing about the creation of a networked person.

We are confronted with changes that have to do with the anthropological features of individuals. We are confronted with a stepwise progression: from being 'scrutinised' by means of video surveillance and biometric technologies, individuals can be 'modified' via the insertion of chips or 'smart' tags in a context that is increasingly turning us precisely into 'networked persons' – persons who are permanently on the net, configured little by little in order to transmit and receive signals that allow tracking and profiling movements, habits, contacts, and thereby modify the meaning and contents of individuals' autonomy.

Technological drifts are therefore taking on especially disquieting features. The purposes of identification, verification, surveillance, security in transactions – may they really justify any use of the human body that is made possible by technological evolution?

These considerations obviously also apply to the cases in which RFID technologies do not result in modifying a person's physicality. To address these issues, one should draw a distinction between the cases in which RFID tags are used as devices directly connected with a given individual (e.g. when they are embedded in an ID card), and the cases in which this link is brought about by the relationships with objects that are, in turn, tagged. In the former case, one has unquestionably to do with situations that are quite similar to those described in respect of body implants, although here the individual has always the option available to get separated from the medium containing the tag and thus escape surveillance – which is unfeasible, or actually much more difficult to do with regard to body implants, including reversible implants. In the latter case, it is necessary to adjust the current legislation on personal data protection by taking account of the pervasive nature of the control and classification that this kind of data collection makes possible – as aptly pointed out in a Working Document adopted by the European Working Party on Data Protection. This implies, on the one hand, the need for re-considering the definition of personal data in order to counter the dangerous trend towards the adoption of formalistic and reductionistic interpretations, which may be prejudicial to the concrete protection of individuals exactly with regard to the applications of RFID technology – and not only this technology. On the other hand, one should seriously take into consideration the risk that standardisation, by allowing access to the data contained in the chip by a plurality of entities and enabling active interventions on such data, might result in controlling and manipulating identity.

This trend was explicitly confirmed by a declaration of the Prime Minister of the United Kingdom on 19 July 2004 to tag and track the five thousand more dangerous English offenders via satellite. The technical difficulties involved in this project have been stressed by many, but it is the symbolic strength of the message that has to be seriously taken into consideration.

Basically it implies a deep change in the legal and social status of a person. The fact of having entirely served the sentence will no longer be enough to be free again. If it is considered as highly probable that a person will commit offences, then that person will lose his freedom of movement and all the relevant forms of individual autonomy, because he will be forced to have an electronic instrument attached to him that will make his location possible at any time. And this tagging of dangerous people can be achieved by putting a microchip under the skin. This would change the nature itself of the body, as by being technologically manipulated it becomes post-human. But can this prospect be considered as compatible with the principle of dignity, which is set forth at the beginning of the Charter of Fundamental Rights of the European Union? Can we accept the Blairian semantic daring move that re-baptised this further version of the 'society of surveillance' as 'the society of respect'? In conclusion: we cannot accept that, networking people, our societies are turned into surveillance, selection, sorting societies; and that via networking and mass control free countries are turned into a nation of suspects.[7]

4. Thus, identity is built up in a scenario where one is increasingly dependent on the outer world – on the manner in which the surrounding environment is built up. One is dependent on other individuals as well as on the things surrounding them or otherwise used to directly change their own bodies. In fact, we are living a true identity revolution, 'the identity (. . .) is in the middle of a period of

extraordinary tumult' (Lasica 2009: 1) in the age of the Web 2.0, of the coming Web 3.0, of the massive profiling, of the new dimensions of the cloud computing and of the autonomic computing. Two changes should be highlighted especially, both being related precisely to the Web 2.0 and 3.0.

Internet 2.0 has become an essential tool for a mass process of socialisation and for the free development of the individual personality. In this perspective, the rights of expression are an essential part of the construction of the person and of its place in the society. The construction of one's identity is becoming increasingly a means to communicate with the rest of the world – to present one's self on the world's stage. This is changing the relationship between public and private sphere, indeed the very concept of privacy.

Privacy has been conceived as an 'exclusion' device – as a tool to fend off the 'unwanted gaze'. However, by analysing the definitions of privacy one can appreciate how privacy has changed over time by giving shape ultimately to a right that is increasingly geared towards enabling the free construction of one's personality – the autonomous building up of one's identity, and the projection of fundamental democratic principles into the private sphere. The initial definition of privacy as the 'right to be let alone' (Warren and Brandeis 1890: 4) has not been done away with; rather, it is now part of a context that has grown out of different contributions. The first real innovation was brought about by Alan Westin, who defined privacy as the right to control how others use the information concerning myself (Westin 1970). Later on, privacy was also regarded as 'the protection of life choices against any form of public control and social stigma' (Friedman 1990: 184) as 'vindication of the boundaries protecting each person's right not to be simplified, objectified, and evaluated out of context' (Rosen 2000: 20) and more directly as 'the freedom from unreasonable constraints on the construction of one's own identity' (Agree and Rotenberg 2001: 7). Since the information flows do not simply contain 'outbound' data – to be kept off others' hands – but also 'inbound' information – on which one might wish to exercise a 'right not to know' – privacy is also to be considered as 'the right to keep control over one's own information and determine the manner of building up one's own private sphere' (Rodotà 1995: 122) 'and as the right to freely choose one's life' (Rigaux 1990: 167). In 2000, the Charter of Fundamental Rights of the EU recognised data protection as an autonomous right. This can be considered the final point of a long evolution, separating privacy and data protection. The evolution is clearly visible by comparing the EU Charter with the provisions made in the 1950 Convention of the Council of Europe. Under Article 8 of the Convention, 'everyone has the right to respect for his private and family life, his home and his correspondence'. Conversely, the Charter draws a distinction between the conventional 'right to respect for his or her private and family life' (Art. 7), which is modelled after the Convention, and 'the right to the protection of personal data' (Art. 8), which becomes thereby a new, autonomous fundamental right. Moreover, Article 8 lays down data processing criteria, expressly envisages access rights, and provides that 'compliance with these rules shall be subject to control by an independent authority'.

The distinction between right to respect for one's private and family life and right to the protection of personal data is more than an empty box. The right to respect for one's private and family life mirrors, first and foremost, an individualistic component: this power basically consists in preventing others from interfering with one's private and family life. In other words, it is a static, negative kind of protection. Conversely, data protection sets out rules on the mechanisms to process data and empowers one to take steps – that is, it is a dynamic kind of protection, which follows a data in all its movements. Additionally, oversight and other powers are not only conferred on the persons concerned (the data subjects), as they are also committed to an independent authority (Article 8.3). Protection is no longer left to data subjects, given that there is a public body that is permanently responsible for it. Thus, it is a redistribution of social and legal powers that is taking shape. It is actually the endpoint of a long evolutionary process experienced by the privacy concept – from its original definition as right to be left alone, up to the right to keep control over one's information and determine how one's private sphere is to be built up.

Furthermore, Article 8 should be put in the broader context of the Charter, which refers to the new rights arising out of scientific and technological innovation. Article 3 deals with the 'right to the integrity of the person', that is, the protection of the *physical* body; Article 8 deals with data protection, that is, the *electronic* body. These provisions are directly related to human dignity, which Article 1 of the Charter declares to be inviolable, as well as to the statement made in the Preamble to the Charter – whereby the Union 'places the individual at the heart of its activities'. Thus, data protection contributes to the 'constitutionalisation of the person' – which can be regarded as one of the most significant achievements not only of the Charter. We are faced with the true re-invention of data protection – not only because it is expressly considered an autonomous, fundamental right, but also because it has turned into an essential tool to freely develop one's personality. Data protection can be considered to sum up a bundle of rights that make up citizenship in the new millennium.

If one probes deeper into the layered safeguards applying to the various categories of personal data, one can appreciate a highly meaningful paradox: indeed, many of the so-called 'sensitive data', especially those concerning opinions, are afforded strong safeguards not so much to better ensure that they are kept confidential, but to enable *public* disclosure of those data without running the risk of discrimination or social stigma. My political opinions or my religious beliefs go hand in hand with and make up my identity only to the extent I can place them outside my private sphere – to the extent I can make them public. The true focus of protection is equality rather than confidentiality.

If identity becomes a relational concept, data protection takes on a different meaning. Social networking, which is the flagship of Web 2.0, mirrors this change in perspective most clearly. You join Facebook because you want to be seen and get a permanent public identity that goes beyond the fifteen minutes of fame Andy Warhol considered to be everyone's right. You feed the 'public' sphere to so that your 'private' can make sense. You exhibit a set of personal data, your electronic body, exactly like one's physical body is exhibited via tattoos, piercing, and other identity signs (Le Breton 2002). Identity becomes communication.

But what happens with this identity that is totally geared to the outer world? It becomes more readily available to data mining activities (Giannotti and Pedreschi 2008, Hildebrandt and Gutwirth 2008) whereupon one might wonder whether social networking also entails an implied consent to the collection of networked data. Or should the principle of purpose specification apply further, whether directly or indirectly, in order to make such data collection legitimate? These questions re-surface if one considers Internet 3.0 – the Internet of Things – and autonomic computing in terms of there being new methods to create and collect personal data.

We are about to experience what an EU research group termed a 'digital tsunami', which might ultimately overthrow the legal tools that safeguard not only the identity, but the very freedom of individuals (The Future Group 2008). We are faced with an in-depth change in societal organisation, whereby the public security criterion is liable to become the sole benchmark.

This objective is stated openly. A document by the EU Council Presidency reads as follows: 'Every object the individual uses, every transaction they make and almost everywhere they go will create a detailed digital record. This will generate a wealth of information for public security organisations, and create huge opportunities for more effective and productive public security efforts'. Furthermore, 'in the near future most objects will generate streams of digital data (. . .) revealing patterns and social behaviours which public security professional can use to present or investigate incidents'. A Statewatch report, *The Shape of Things to Come* (the same title of a 1933 novel by H. G. Wells) (Bunyan September 2008) shows how the EU has substituted the concept that data relating to EU citizens should in principle be kept private from State agencies, in favour of the principle that the State should have access to every detail about our private lives. In this scenario, data protection and judicial scrutiny of police surveillance are perceived by the EU as 'obstacles' to efficient law enforcement cooperation. This implies that European governments and EU policy-makers are pursuing unfettered powers to access and gather masses of personal data on the everyday life of everyone, on the ground that we can all be safe and secured from perceived 'threats'.

The criticisms by Statewatch are levelled against a specific feature of the digital tsunami – the growing use of the public security argument to downsize freedoms and rights and turn our societal organisation from a society of free individuals into a 'nation of suspects'. This is unquestionably a key issue because it has to do with the change in the relationship between citizens and State; more specifically, it is a violation of the undertaking made by the State vis-à-vis every individual that their data will be used selectively in compliance with such principles as data minimisation, purpose specification, proportionality, and relevance. In this manner, some of these principles underlying the system of personal data protection are being slowly eroded. This applies, first and foremost, to the purpose specification principle and the principle concerning separation between the data processed by public bodies and those processed by private entities. The only principle to be referred to becomes the principle of availability, with a view to improving the exchange and use by law enforcement agencies. The multi-functionality criterion is increasingly applied, at times under the pressure exerted by institutional agencies. Data collected for a given purpose are made available for different purposes, which are considered to be as important as those for which the collection had been undertaken. Data processed by a given agency are made available to different agencies. It means that individuals are more and more transparent and that public bodies are more and more out of any political and legal control. It implies a re-distribution of political and social powers.

This means that the so-called digital tsunami should also be evaluated from different standpoints – starting exactly from the identity perspective. The full availability of all personal data to public agencies brings about a veritable transfer of identity into the hands of such agencies, which can actually rely on information that is unknown to the given data subject. This phenomenon is bound to take on increasing importance in view of the increasing amount of object-generated information. One comes face to face, in this manner, with one of the key features of data protection – the right of access, meaning everyone's unconditional power to know who holds what data concerning them and how such data is used. This can be the stepping stone to start re-constructing one's identity, by having any data that are untrue, obtained unlawfully and/or kept beyond the allotted time cancelled or erased; by having any data that are inaccurate rectified; and by supplementing any incomplete data. However, this has turned by now into a never-ending story – a bottomless chasm, because never does the recording of every trace we leave come to an end. 'Know thyself' is no longer a precept that requires us to only probe into ourselves. Indeed, it relies on the assumption that one should manage to get back to different sources in order to establish not so much what the others know about us, but who we are in the electronic dimension – where a major portion of our lives are nowadays to be found. It has been said that identity is no more 'what you say you are', but 'what Google says you are'.

We have to do with issues that relate to autonomy and the right to freely develop one's own personality. Everyone's freedom to know and build up their own selves is being increasingly constrained, whilst it is increasingly easier for others to become the lords and masters of our lives.

5. Thus, it is not enough to remark that we are living by now in a 'networked public sphere', to quote Yochai Benkler (2006). It is necessary to consider how this public sphere is being created, by whom it is created, and how the public sphere in question is being shaped on account of these changes. The stepwise descent into a smart environment populated by intelligent things is giving rise to yet another shift, which goes beyond what has brought about the stepwise separation/opposition between one's self and the others as related to building up one's own identity. Indeed, there is a widening gap between individuals and machines due to the growing autonomy of the latter as mirrored by the so-called autonomic computing. The power to set the boundaries of human beings and their identity is

increasingly shifted from the realm of man's appreciation to that of automated decisions.

An analysis of the above issues raises several questions. One can start from Article 15 of EU Directive 95/46 on the protection of individuals with regard to the processing of personal data and on the free movement of such data. This Article, at point (1), provides that 'member States shall grant the right to every person not to be subject to a decision which produces legal effects regarding him or significantly affects him and which is based solely on automated process of data intended to evaluate certain personal aspects relating to him, such as performance at work, creditworthiness, reliability, etc.'. For the sake of simplification, one might argue that this is a general provision on the allocation of decision-making powers in the digital world.

However, the symbolic as well as practical import of this provision is markedly reduced by the restrictive interpretation espoused by several domestic laws and, in particular, by the increasingly widespread, sophisticated application of profiling techniques – which have changed the very meaning of 'decision'. The studies on data mining and profiling have highlighted the basically regulatory importance attached to categorisation, which is often socially more binding than legally binding decisions. This point is actually made in the EU directive, which refers to decisions 'significantly affecting' persons. Profiling results in social sorting and accordingly brings about the risks of social stigma and exclusion. It is no chance that the definition of privacy had already emphasised the risk of social stigma well in advance of the coming of the age of profiling.

The creation of this new environment results in changes in individuals' behaviour; these changes have often been described and consist in self-restraint, 'spontaneous' normalisation, and the a priori adoption of mainstream behaviour. Profiling mirrors the modelling of society, which gives rise to mainstream behaviour rather than normalcy – which is actually no news to the scholars of cultural models, because the influence of such models is not a function of their formally binding value, but of them being regarded as a necessary step in view of social acceptance at the most diverse levels. This effect is enhanced further by data mining and profiling, because models are customised and linked to single individuals – ultimately, they are used in a targeted, selective manner. Social acceptance is shaped thereby as a kind of 'compulsory' identity.

The possibility to escape from this mandatory scenario by relying on the system introduced by directive {95/46} is jeopardised further by the considerations put forward to criticise the view that exclusively automated decisions are unacceptable. It is argued that the human presence – considered to be a fundamental component of any legitimate decision-making – features from the start in this context: 'humans will write the software, shape the database parameters and decide on the kinds of matches that count' (Schwartz et al. 2008: 282). This argument is supported further by the consideration that automated decision-making processes are increasingly similar to human decision-making. It is no chance that the autonomic computing paradigm has been inspired by the human autonomic nervous system.

'Its overarching goal is to realize computer and software systems and applications that can manage themselves in accordance with high-level guidance from humans' (Parashar and Hariri 2004: 247). If what is artificial is increasingly modelled after nature, there is no longer any reason for the ban set forth in Article 15 of directive 95/46. The severance from the human factor is seemingly complete.

In the light of this blunt elimination of any boundaries between human and artificial processes, it should be pointed out that the staunchest supporters of the above stance are those who claim that markets and security should have the upper hand. Monitoring individuals and turning them into mere consumers are regarded as priority objectives that allow for the use of any tools. This impacts directly on identity, which is increasingly built up via external entities – and the interests vested in such entities may be completely different from those vested in the given individuals, who are deprived accordingly of any opportunity both for exercising their self-governance powers and for controlling who has got hold of their identities.

Can one re-appropriate at least some measure of control by relying, first and foremost, on the guidance set forth in directive {95/46}? There are three points to be made in this connection. Firstly, it is necessary to uphold the principle whereby a wholly automated decision should never replace a decision that involves some sort of human involvement. Secondly, one should consider the access rights vested in every data subject under Article 12(a) of the directive, in particular with a view to knowing 'the logic involved in any automatic processing of data concerning him, at least in the case of the automated decisions referred to in Article 15(1)'. The emphasis put on logic is especially important because it is linked to another key feature, that is, the contribution given by technology designers (Rouvroy 2008: 44) and the circumstance that digital technologies are built around their own 'code' (Lessig 1999) – which means that the control issue is also to be taken into account. Furthermore, since the limitations and constraints applying to subject access are widely known (Hildebrandt and Gutwirth 2008: 303–37) – and this is the third point to be made – one should envisage access by collective entities acting on a data subject's instructions, which is already the case with some domestic laws. This solution – which is reminiscent of the approach that has emerged from the history of trade unions – would help reduce the power unbalance of the individual stakeholders, bring about enhanced transparency, and above all initiate wide-ranging control processes based on societal self-organisation schemes.

In this perspective, four more issues must be taken into account, giving room to a new set of rights. The first one looks at the 'anonymity' (alias, pseudonym) and to the encryption as tools that make possible not to be identified and, consequently, not to be profiled, with a dichotomy between online persona and real-world identity. In a 25 March 2010 decision the Israeli Supreme Court stated that the right of anonymity is an essential part of the Internet culture, because it offers a safe outlet for the user to experiment with novel ideas, express unorthodox political views or criticise corporate or individual behaviour without fear of intimidation or reprisal.

The second looks to 'the right to make silent the chip', immediately referred to the RFID technologies, but that can be extended to the many devices making possible the multifarious forms of distance control (localisation and so on). This right empowers individuals to cut the connection with technological apparatus collecting informations on them.

The third underlies the necessity of new guarantees and 'sunset rules' on data retention. In a decision of 2 March 2010 the German Constitutional Court has declared incompatible with the constitutional legal identity of the Federal Republic of Germany the Directive 2006/24/ES.[8] It stated that 'the protection of communication does not include only the content but also the secrecy of the circumstances of the communication, including especially if, when and how many times did some person (. . .) contact another or tried that', because 'the evaluation of these data makes it possible to make conclusions about hidden layers of a person's private life and gives under certain circumstances a picture of detailed personality and movement profiles'.

The fourth is connected directly with cloud computing. Cloud computing is a shorthand for defining a vast, always-on, accessible, broadband-enabled next-generation Internet. This new dimension must be looked at in connection with the changes of the online identity produced by the social Web. 'Blogging became a mainstream activity, and with it came a different mind set. With a few exceptions, bloggers found the need to stand behind their words. They needed to tie their online musing to their real lives. Authenticity and transparency – not imagination and anonymity – became the cardinal rules of the blogsphere' (Lasica 2009: 16). This conclusion could be criticised, but it is true that with the rise of You Tube, of the video-sharing sites, of the social networking sites such as MySpace and Facebook the situation has changed, and Facebook became the first Web service requiring validated identities, even if it is not difficult to create false identities, becoming the largest platform of the cloud computing era. It implies that the personal data posted in Facebook need a fresh approach, because the traditional protection provided by the principle of the consent could not work, due to the fact that the posting is voluntary. Taking into account this novelty, it has been suggested to make reference to the purpose specification principle, so that the personal data made public for a merely social interaction with other people cannot be accessed and treated for different finalities (marketing, control).

The identity in the cloud has suggested a new approach to the identity itself in the social context, going toward a 'user-centric open identity network'. The idea is

> an identity system that is scalable (so it works everywhere), user-centric (serving your interest, instead of something done to you by outside interests) and, importantly, customizable. This new system would recognize that each of us has multiple identities. We will be able to spoon out bits and pieces of our identity, depending on the social or business context we find ourselves in (. . .) You could separate your identity into discrete units and assign different access permission depending on your role in a given situation. You could create a business profile, a health care profile, a friend profile, a mom or singles

profile, a virtual profile and so on (. . .) Few identity software developers expect that most people want to manage their own identities.

(Lasica 2009: 17–18)

Can we look at this suggestion as a road to the reconquest of individual power on identity?

Looking at the multiple faces of the identity, we can escape the risks of 'obsession' of the 'unique' identity (Remotti 2010) and outline different scenarios for human identity. For instance, it has been suggested that we can have our present selves (Someone); a hedonist depersonalised version (No one); a socially oriented self (Anyone); an autonomous creative individual (Eureka). Technology shapes a world in which all four persons can be developed in an integrated portfolio (Greenfield 2008). It means that identity spans many contexts and purposes and that identity management has become a central issue (Wessels 2010).

6. Autonomic computing should be regarded in the perspective of this force field. It is a fact of life that should be understood in cultural terms and requires mechanisms of societal control, which need not rely on laws only. It is unquestionable that we are faced with the re-definition of the whole context applying to the relationship between identity and autonomy, which is bound to impact on the meaning and import of these two concepts. In fact, autonomic computing might mark the ultimate separation between autonomy and identity. Identity is becoming objective via mechanisms that do not rely on awareness of one's own self; it is turning into a functional replacement for autonomy – at least to the extent an adaptive scheme is built out of an identity that was 'captured' at a given time, with all the respective features and needs, whereupon that identity is committed to self-management systems that can provide responses and meet the requirements arising out of the given circumstances. The construction of this 'adaptive' identity might be regarded as a process that originates exactly from the freezing of that identity and keeps on adapting it to the relevant environment without any decisions being made and/or any awareness being developed by the individual. This is made possible by the unrelenting collection of information yielding statistical estimates that can accordingly anticipate/implement what the individual data subjects would have decided in the given circumstances. The possibility of conscious interventions by the individuals could be totally excluded, making impossible their participation by default too. The construction of identity depends on algorithms, and we must be aware of the role played by mathematic models and algorithms as key elements of an economic organisation that has produced the financial crisis.

In fact, 'the environment can act upon the user's behalf without conscious mediation. It can extrapolate behavioural characteristics and generate pro-active responses' (Aarts and de Ruyter 2009: 8). One might argue that we are faced with the separation between identity and intentionality, which may give rise to unaccountability, discourage the propensity to change, and jeopardise a vigilant approach to the governance of one's self. Indeed, one should wonder whether the activities performed via autonomic computing are a projection from the past rather than the anticipation of future events (Hildebrandt and Gutwirth 2008: 4). We are risking a final expropriation of the identity as a product of the person's autonomy.

7. Autonomic computing, which is not expressed solely by the specific functions described above, is setting the cultural, political and institutional agenda. It is bringing about an information collection mechanism that – as well as being wide-ranging and pervasive – is not static, but rather intrinsically dynamic. This means that it produces effects without any mediation being required, without having to make the information available to other entities or else use it for subsequent processing operations. Obviously, given the manner in which personal data are collected, opportunities also arise for such data to be used further, which increases not only the chances to automatically meet the given requirements, but also the overall transparency of individuals. And it means that, at the same time, not only a new 'inner' space of the person, but an 'outer' space is currently being shaped.

Given this context, one is bound to turn once again to personal data protection seen not only as 'a necessary utopia' (Simitis 2005), but also as a way to ensure the freedom of individuals and conditions for the democratic exercise of powers. Indeed, the technological changes impacting on social organisation do not only give rise to unbalances in the distribution and practice of power – they also bring about a societal gap between increasingly transparent individuals and increasingly opaque, unbridled powers.

Especially after 9/11, the boundless collection of personal data has been used as an indispensable tool to counter terrorism, based on the argument that who has nothing to hide has nothing to fear from whatever collection of information concerning them. Still, we should not forget that the 'glass man' is a Nazi, totalitarian metaphor. Who wishes to retain a private, confidential sphere is automatically labelled as a 'bad citizen' and can be the target of oppression.

Thus, the starting point for looking at data protection in a new perspective can be said to consist in highlighting the mechanisms required to counter the coming of the digital tsunami. However, in the age of networked persons, the Internet of Things, and autonomic computing, this means to be also afforded a general 'freedom to disconnect', to have the right 'to make silent the chip', as already remarked in connection with the purchase of goods containing a chip that can allow information on the purchaser to be subsequently acquired.

8. The conditions applying to the electronic body and digital identity are similar, at least in part, to those applying to the physical body, given that the insertion of electronic devices into the human body or its dependence by the way some things are working can hamper one's autonomy to a greater or lesser extent and make one dependent on the outer world. A precondition for the insertion of a chip to be lawful consists first and foremost in the reversibility of the implant – which ensures that the individual can retain governance over their own body.

It is exactly in looking at the physical dimension of the individual that one cannot help raising questions concerning identity. 'Is a hybrid bionic system a person, an entity one can attribute rights and duties on that account? (. . .) Is the human user/

component of a hybrid bionic system the same person before and after being so interfaced with artificial devices?' (Tamburrini 2008: 451).

Identity is therefore shifting from a synchronic to a diachronic dimension, which also holds true if the individual is digitised and becomes part of an electronic network. It is the time-honoured topos of Theseus' ship, which can be conjured up nowadays as a token that can help us better realise how necessary it is to replace a static concept of identity by a dynamic one – whereby identity becomes as varying as the individual it applies to whilst its epistemological nature is unrelentingly reshaped. Subject is no longer compact, unified, neatly defined entity. It is enigma (Castoriadis 1990) rather than problem. It is becoming nomadic (Braidotti 1994) – which mirrors the fragmented, mobile reality. It is no harbour – rather, it is process. And identity may be equated to the many 'windows' that open on a screen. 'These windows have become a powerful metaphor to conceive of the self as a multiple, distributed system' (Turkle 1995: 14). However, this multiplication of identities does not only result from the activities of several entities looking at the same individual from multifarious viewpoints. In cyberspace, the data subject can continuously take on different identities in order to better communicate and get rid of constraints that would hamper the free development of their personality. A necessary assumption in this scenario is the right to anonymity: even if it has been harshly questioned in the age of the endless war on terror, however, it still remains a key component of citizenship in the new millennium. Indeed, there is no conflict between the permanent vindication of this right, which entails opaqueness, and the rampant social networking that is the utmost manifestation of transparency. There are increased options available to build up identity, and people must be in a position to make use of all those options.

We are faced ultimately with the concept of identity as a process, and this is clearly shown by digital identity management systems: these systems should arguably meet three core privacy requirements. The system must (1) make data flows explicit and subject to data owners' control; (2) support data minimisation by disclosing no more data are needed in a given context; and (3) impose limits on linkability (Schwartz et al. 2008). Still, these requirements, which are compounded of legal norms and privacy by design, should not be regarded as the ultimate solution – rather, they are markers to be used in order to enhance societal awareness of identity-related issues.

Notes

1 The title of this paper mimics that one of a well-known John Steinbeck novel, 'Of Mice and Men' (1937), a story of an unconscious destructive power that can be stopped only through the destruction of that power itself. Violence, public or hidden, against violence? We must avoid any aggressive attitude. To mimic William Shakespeare (King Lear, act V, scene II, 'Ripeness is all') we could say 'Consciousness is all'.
2 M. de Montaigne, *Essais* (1595), Livre III, Chap. III, De trois commerces (Garnier, Paris, 1962, t. II, p. 238).
3 The Works of Francis Bacon, vol. III. Magnalia Naturae (1627) appended at New Atlantis, Longman, London, 1857, pp. 167–8.

4 BVerfGE 1 BvR 357/05, available www.bverfg.de/entscheidungen/rs20060215_1bvr 035705en.html.
5 BVerfGE 1 BvR {370/07}, 1 BvR 595/07, available www.bundesverfassungsgericht.de/ en/decisions/rs20080227_1bvr037007en.html.
6 15 December 1983 BVerfG 65, 1 (the census decision) and BVerfG NJW 2004, 999.
7 It is noteworthy that in the programme of the new UK Cabinet there is a chapter on 'Civil Liberties' proposing a completely different approach to the use of surveillance technologies.
8 BVerfG, 1 BvR {256/08}, available www.bundesverfassungsgericht.de/entscheidungen/ rs20100302_1bvr025608.html.

References

Aarts, E. and de Ruyter, B. (2009) 'New research perspectives on Ambient Intelligence', Journal of Ambient Intelligence and Smart Environments, 1: 5–14.

Agree, P.E. and Rotenberg, M. (2001) Technology and Privacy. The New Landscape, Cambridge, MA: MIT Press.

Benkler, Y. (2006) The Wealth of Networks: How Social Production Transforms Markets and Freedom, New Haven: Yale University Press.

Braidotti, R. (1994) Nomadic Subject: Embodiment and Sexual Difference in Contemporary Feminist Theory, New York: Columbia University Press.

Bunyan T. (September 2008) The Shape of Things to Come, London: Statewatch.

Castoriadis C. (1990) 'L'état du sujet aujourd'hui', in Le monde morcelé. Les Carrefours du Labirinthe, III, Paris: Seuil: 189–225.

D'Holbach P.H.T. (1990) Système de la nature (1770), 2 vols, Paris: Fayard.

de La Mettrie, J.O. (1987) L'homme machine (1748), in {OElig}uvres philosophiques, 2 vols, Paris: Fayard.

Friedman L. M. (1990) The Republic of Choice. Law, Authority and Choice, Cambridge, MA: Harvard University Press.

Greenfield, S. (2008) ID: The Quest for Meaning in the 21st Century, London: Sceptre.

Giannotti F. and Pedreschi D. (eds) (2008) Mobility, Data Mining and Privacy. Geographic Knowledge Discovery, Berlin-Heidelberg: Springer.

Hildebrandt, M. and Gutwirth, S. (2008) 'General Introduction and Overview', in M. Hildebrandt and S. Gutwirth (eds), Profiling the European Citizen. Cross-Disciplinary Perspectives, Dordrecht: Springer, 1–13.

Lasica J.D. (2009) Identity in the Age of Cloud Computing-The next-generation Internet's impact on business governance and social interaction, Washington D.C.: The Aspen Institute.

Lessig L. (1999) Code and Other Laws of the Cyberspace, New York: Basic Books.

Le Breton, D. (2002) Signes d'identité. Tatouages, piercings et autres marques corporelles, Paris: Metailé.

Losano, M.G. (1990) Storie di automi. Dall'antica Grecia alla belle époque, Torino: Einaudi.

Mullins, A. (2008) Declaration found in a press release on the Court of Arbitration for Sport decision on Pistorius case, May 16, 2008.

Parashar, M. and Hariri, S. (2004) 'Autonomic Computing: An Overview', in J.-P. Banatre et al. (eds), Unconventional Programming Paradigms 2004, LNCS 3566, Berlin-Heidelberg: Springer: 247–59.

Remotti, F. (2010) L'ossessione identitaria, Bologna: Il Mulino.

Rigaux, F. (1990) *La protection de la vie privée et des autres biens de la personnalité*, Bruxelles-Paris: Bruylant.

Rodotà, S. (1995) *Tecnologie e diritti*, Bologna: Il Mulino.

Rosen, J. (2000) *The Unwanted Gaze. The Destruction of Privacy in America*, New York: Random House.

Rouvroy, A. (2008) 'Privacy, Data Protection, and the Unprecedented Challenges of the Ambient Intelligence', (2) *Studies in Ethics, Law, and Technology*, 1, 1–51.

Sartre J. P. (1954) *Réflexions sur la question juive*, Paris: Gallimard.

Schwartz, P.M., Lee, R.D. and Rubinstein, I. (2008) *Data Mining and Internet Profiling: Emerging Regulatory and Technological Approaches*, Berkeley Center for Law and Technology, *Paper 50*.

Simitis S. (2005)'Datenschtz – eine notwendige Utopie', in *Summa. Dieter Simon zum 70. Geburtstag*, Frankfurt am Main: Klostermann.

Sini, C. (2009) *L'uomo, la macchina, l'automa*, Torino: Bollati Boringhieri.

Tamburrini, F.L. G. (2008) 'Ethical monitoring of brain-machine interfaces. A note on personal identity and autonomy', (22) *Artificial Intelligence & Society*, 3: 449–60.

The Future Group (June 2008) *Freedom, Security, Privacy – European Home Affairs in an Open World*.

Turkle, S. (1995) *Life on the Screen. Identity in the Age of the Internet*, New York: Simon & Schuster.

Warren, S. and Brandeis, L.D. (1890) (4) 'The Right to Privacy', *Harvard Law Review*, 5: 193–220.

Wessels, B. (2010) *Exploring the Practises of Identity and Privacy in Digital Communication.* www.iseing.org/tgovwebsite/tgov20Acceptedpapers/C19.pdf

Westin, A. (1970) *Privacy and Freedom*, New York: Atheneum.

Chapter 11

'The BPI Nexus'

A philosophical echo to Stefano Rodotà's 'Of Machines and Men'

Paul Mathias

> Usez de votre moi pendant que vous le tenez.
>
> Victor Hugo, *Les Misérables*

A computed world

Computers and communication devices – mobile phones, personal digital assistants and lately iPods as well as other portable and geek-oriented paraphernalia – are not merely instruments for connecting people for work and pleasure. This is why they shouldn't be thought of on just a technical level, and their use is not all about skilfulness and appropriateness. Indeed, not only do they allow us to interconnect for business and leisure purposes, but they are also connected among themselves to form a sort of information nebula in ways we rarely think about and are even difficult to imagine. Digital artefacts form a 'universe' of their own. This is not to say they constitute a technical body animated by a 'digital intelligence'. But it is quite obvious that they are able to calculate data and produce results beyond their users' scope and consciousness. Individuals or groups, institutions and corporations praise their efficiency but have little or no consideration for the type of reality machines have ended up building and composing for us to live in and with.

For we certainly live in a realm of calculations: calculations made by computers generating the texts and images to which we are continuously exposed, calculations made by routers dispatching information among any of the devices connected to the Internet and other networks, calculations made by computer grids about calculations made by computers and routers. The amazing quantities of data generated by our computing machines, whatever their location, size and speed, transcend the possibilities or capacities of our minds. Not that we are unable to use such data. In fact many machines and programs are used to efficiently dig data out of repositories and combine them into relevant forms of information and knowledge – *infoledge*.[1] This is exactly what happens when we make use of a search engine and peruse information to discover mixed facts, events, ideas, theories. Machines may not be clever enough to answer all of our questions, but they certainly are dumb enough to cycle through tremendous amounts of data in order to meet the needs of our will and reflection – or desires, for that matter. And the paradox is that calculations are so insanely numerous that they eventually meet our demands without us really understanding what is actually happening, how it is happening and the very status of such a coincidence: is it plain chance, pre-formed knowledge, self-generated information, what is nowadays called 'serendipity' – or is it the mere product of one or more proprietary thus hidden algorithms?

Similarly, no part or aspect of our economic, social, political or private existence is free from the calculations of machines we rely on, willingly or not. The use of an everyday life apparatus such as a telephone, a computer, any means of transportation, or even appliances, rely upon digital systems managing their everyday use – if only the industrial grid allowing for the distribution of electrical energy among houses and offices. Many if not all our 'voluntary' actions (De Mul 2009) are essentially dependent on the intricacy of numerous technological systems constituting, as a whole, the contemporary ecosystem of human life. This means the symbolic, religious and institutional milieu within which individuals and societies prospered in the past has been doubled or even tripled by the development of labyrinthine digital as well as energy grids. Whatever our beliefs, they are carried out, so to say, by the techno-industrial pattern onto which they are currently being transcribed in the course of our post-industrialisation, in forms as varied as published text and images, networked transmission trails, broadcasting environments and the like. Now obviously, the contemporary digitalisation of life does not make us 'cognitive supermen'. Most certainly, it makes us *trivial* hybrids: an ordinary citizen filing online his or her tax forms to the IRS, a man or a woman flying across Europe to meet a partner, a child riding the TGV through France from Lille to Marseille, a truck driver having his route tracked for security purposes, all following a trail of endless enhancements of their daily lives as well as intellectual and practical abilities. The contemporary digitalisation of life just makes contemporary life possible and real.

Quite evidently 'existence' and 'life' are not mere abstractions. If we are to try and interpret what they mean, a 'flesh and blood' frame of reference is conceptually unavoidable. This frame of reference is nevertheless problematic because of a renewed complexity in our lives that are inundated with digital instruments and multiple forms of waves, both aerial and terrestrial: fibre optics, WIFI antennas, Bluetooth connections, etc. What we are – a body, a person and her identity or identities – forms a very intricate phenomenological complex: the 'Body Person Identity compendium' hereafter named *The BPI nexus*. While the correlation between body, person and identity is not new to philosophy,[2] the *digital hybridisation* of our lives changes radically the patterns for our reflecting upon it. Traditionally the symbolic dimension attached to the 'BPI nexus' was related to our own building up of ourselves as well as our social inscription in axiological moulds such as standing, dressing, behaving (Elias 1969, Mauss 2006). Academic studies about tattoos and religious or other forms of scarification have also shown that the body may stand as a picture for the person and her 'true' identity (DeMello 2000, Renaut 2004). In this regard, contemporary stadiums and 'gyms' are quite similar to the Greek *palestra*: places to develop one's muscular architecture and social acceptability.

But these are not the only things that effect what we call the 'BPI nexus'.

Perfecting technology, miniaturising computers, ramifying the complexities of the man/machine compendium have provoked a profound mutation in our representation of ourselves and perhaps even in the actual 'substance' of our being. The whole process of postmodernity is not anymore about writing our own singular or social self within our muscles or upon our skin; it is about *mutating* and letting our very *nature* glide into a hybrid existence. Our bodily capacities – what we are capable of or allowed to do – as well as our personality and identity are at least partly redefined by the pervasiveness of many combined digital instruments. *Machines make us*, obviously, not in the sense that we all more or less carry technological grafts, but in the sense that we can neither do without nor away with such extensions. Our being has thus become 'naturally' digital. What we mean by this is that digital artefacts are less and less alternative forms of reflection and will, and more and more extensions of our intellectual and practical actions – like when the braking system of a car 'chooses' how to 'interpret' a driver's pressure on the pedal. Yet what exactly is 'making' for a machine? And what is 'being made' for a person? Should we say that such a reciprocal process translates our very *being-there* into virtual terms of calculability? Is this to say that whatever we do and whenever we do it, practically or intellectually, this will at least partly be the resulting process of unrepresentable amounts of calculations that we know very little about? Is this to say that our very existence should be regarded as the result of various processes of virtualisation?

To be able to answer these questions, we need to try and figure out the meaning of our *technological being* and measure up the conceptual spectrum of its existential horizon. Reflecting anew upon the 'BPI nexus' means reaching a greater consciousness of the true frontiers of our existence. 'Thinking while living' no longer accurately describes us. The conceptual depiction of our lives' digitalisation is among the most challenging meditations. Not just for the artist and the novelist, but for the philosopher and the lawyer, both of whom are interested in the technical and judicial means for our *non-alienation* to – if not *liberation* from – the digital artefacts we produce and the technology we let invade us. Our power to produce automated objects seems to have grown into the power of producing human beings very close to being automated objects themselves. The *natural* hybridisation of Man was fantasised by Diderot in his description of a 'tireless race of goat-men'.[3] We have now reached the age of a *technological* hybridisation of Man and new forms of confounding 'humanity' with 'thingness'. Brillo boxes are production units that are labelled and distributed to warehouses and from there to retailer stores. So are men, at least some of them: sometimes children, more often criminals or elderly people suffering from heavy neurological diseases. Which is not only a matter of technological 'enhancement' but new ways of building up social relations and keeping them ordinate.

Obviously, there is nothing new to the *contiguity* between 'Man' and 'Machine'. With stone or wood, metal and energy, the world has always been built with tools, instruments and more recently machines, and mankind has always lived and developed itself within complex technological environments. What is truly new, though, is the present *continuity* that characterises the man/machine compendium. 'Continuity' refers to an internalisation of the (miniaturised) machine within the human body and the externalisation of the (computed) human will and understanding onto the computer assisting a person – a pilot in his or her jet fighter for instance or even a writer using the 'suggestions' made by his or her favourite word processor's thesaurus. This written contribution – and echo – to Stefano Rodotà's 'Of Machines and Men' is an attempt to theorise this 'continuity' as a key feature of what has been named 'the BPI nexus', which is a construction perpetually in the making. This paper will argue that technology and the internalisation of artefacts helps construct and re-assert the very reality we have very good reasons to call our 'self'. As a hybrid, the body is only an element of a much larger process endlessly constructing and (re-)defining us as persons, private and public, social or solitary. A process that is sublimated by the narratives of our postmodernity deployed not only as theories, fantasies, art and philosophy, but also as technical instruments and languages allowing for their use. Not only can we say what and who we are as autonomous persons, but machines are from now on meant to be aspects of our singular or multiple identities. All of us are imperceptibly absorbed into mechanical and automated processes and becoming instruments identified with various predetermined functions: being a soldier, a teacher, a physician or a citizen.

Body building: otherness renewed

Looking back to the mid 70s, one might recall the famous *Six Million Dollar Man*, a TV serial starring actor Lee Majors as Col. Steve Austin. Austin portrays an ex-astronaut who after crashing his capsule is rebuilt as a 'bionic' man. Neither a man nor a machine entirely, but a man-machine, his organs and members having been 'enhanced', he is able to see and hear farther, to run faster and act more quickly than any naturally endowed athletic man or trained military personnel. Mechanical as well as digital prosthetics make Col. Austin a human being beyond humanity, more efficient and successful, wiser, not far from immortal – a tale of post-humanity imagined in the Hollywood workshops of our contemporary world.

One interesting thing about the *Six Million Dollar Man* is that it shows how tampering with a man's organs does not simply lead to repairing an ailing part of his body. It has largely to do with creating new horizons for the individual, thus deflecting one's identity or idea of one's self. This is precisely why we need to reflect upon the technological enhancement of Man, not enhancement by means of instruments only but mainly through the *internal* transformation of his organs and body parts. Enhancing an aeroplane's ability to fly faster and higher is one thing; enhancing a man's capacities to climb up hills and mountains on a bike is something radically different. Not only is it intrinsically more complex but it certainly has to do with that man's 'being', not just his or her physical capabilities. It has to do with his or her body, personality and identity. Is Col. Steve Austin really the *same* man before and after his crash and techno-medical treatment?[4] What about the biker and the champion? It is not enough to observe that their appearance and voice have not changed. And the point is not that they can be used differently and more adequately to their more or less new capacities but, as a matter of fact, that they can be *utilised* as *reduced* to their more or less extraordinary properties and powers. As shaped in Hollywood or by the contemporary pharmaco-medical industry, the 'overman' does not resemble a man anymore and should more accurately be likened to an animated instrument. Similar to Aristotle's slave (*Politics:* I, 2), his physical as well as intellectual faculties are almost exclusively alienated to the requirements of his missions and environment. In the case of the *Six Million Dollar Man* we witness a tale of transformation and exploitation[5] – a tale of *alienation* of an individual in the form of body enhancement, personality vectorialisation and identity redefinition. 'Body enhancement' leads to 'personality vectorialisation', the modification of organs or members being technologically pre-defined as aiming towards a military, not a meaningless purpose; such a process results in 'identity redefinition' since it is implied that the man-machine's social space and existence are strictly confined within the perimeter of his technological determinations – the individual being characterised by his usability, disposability, if not expendability.

The popular culture character Steve Austin outlines the fact that body integrity is a very important *moral* and *social* issue. It is nowadays even more intricate due to a conflict between our physical and our electronic or digital integrity. We have learned to deal with the former, as it is generally if not universally accepted that one may not be legitimately submitted to gratuitous violence, amputated, enslaved, forced to act against one's will. The latter may still be – at least partly – strange to our reflective and critical patterns. To use Stefano Rodotà's words, 'the electronic body' certainly has to do with digital implants and 'tools that entail [its] manipu-lation' – which all conflict with our idea of 'human dignity'. Yet the frontiers between the electronic and the physical, as well as their *metaphysical* consequences on our comprehension of the body *as such* are not clearly established.

To start from a radically pre-digital point of view, it is important to understand that our body is not merely a body. First of all it is not *a* body among others, like wood, metal, even organic substances. It is *our* body and a matter of personality and identity. A big man or woman does not move and act like a frail one, or the supposedly 'beautiful' with the same discretion as the supposedly 'ugly'. Social stigma affects us in our physical being and make us be however our body seems to imply that we are. Yet this does not mean we *are* our body, even though our body may express something that we are when it comes to various pathologies.[6] Nor does it mean that we *have* or possess or carry a body, as if it were some sort of a property, thus disposable and/or negotiable. To praise human dignity is to consider the body not as just a material support for activities of another nature, for example intellectual or moral. Thus we might say both that we are and have a body or that we neither are nor have a body. In fact our body means *otherness*. It certainly is the closest thing to what we name and describe as 'our being'. And indeed whoever and whatever we are, intimately or socially, exudes from our body's pores as a varied construction as well as materialisation of our very own 'self': we are incarnated by the shape of our body, but also by our voice, our gestures, even the slightest movement of our hands or lips. Yet no one would dare reduce oneself to the state of one's body. I am healthy without being the health of my body, and if I am young and beautiful, woe betide me if I think I am my youth and my beauty!

Under these circumstances, reflecting upon our body's digital extensions should not convey the idea that we simply switch from a pattern of autonomy – the identity of a person's incarnation – to a pattern of heteronomy – machines and things taking over the monitoring of our own 'self', first of all because our body does not carry the meaning of 'interiority', 'self determination' or 'autonomy'. In a sense, to the extent that it has a life of its own, our body has always meant otherness to our own self. As a consequence, what is truly and newly problematic is the imperceptible switch from one scheme of heteronomy to another – two different ways in which we come to terms with our own otherness. In the first case, otherness is experienced as our body's spontaneous movements of behaviour and we are confronted by our body alone.[7] In the other, we need to face our 'augmented' and restructured body, its hybridised nature – and in return how technology itself is made nature.

Within the naturalistic frame, 'self knowledge' is considered as related to the physical as well as intellectual building of the self. *Working* is the centre of gravity of our personal architecture. Reading, swimming or ploughing are but forms of working, muscularly or otherwise. They represent a dual process of projection of the inner self towards Nature, and of integration of the necessities of the objective world into the patterns of the mind and the self. By acting, we externalise thoughts and ideals, clearly determined or not; and by experiencing the world through acting, we internalise its requirements and learn to form our own selves.[8] Our whole apprehension of the body and its symbolic meaning, mainly its dignity, comes from intellectual patterns more or less akin to this one. Connecting our personality and identity to our body structure and behaviour is not simply *relying* on its materiality; rather it is following its own history and series of transformations, its own otherness. We either fail to be who and where we think we are – by imagining or pretending we can achieve things beyond our true scope – or exceed our own possibilities by being creative and overcoming our own existential context. In the naturalistic frame, the 'subject' to 'object' relation does not cover one of 'self' to 'body'. They all are intrinsically one and form the singular history – and possibly tragedy – of each human being's life and destiny.

It remains that Stefano Rodotà is absolutely right to focus on a new form of 'self knowledge' related to the externalisation of its sources. In the henceforth digital frame of heteronomy, our history is not anymore that of our own possibilities and their inclusion in the naturally objective world. Our history itself becomes digital mainly because machines gather, save and preserve data concerning our lives, actions, beliefs, our past and our present, even our expectations thus our future. Of course it is all still about building our 'self'. Only the operation of building the being that we tend to be is not anymore exclusively related to our own power but *diluted* into a myriad of digital processes we have almost no idea of. There is a new dimension to the simple fact of being which is due to the 'compenetration' of the man and the machine, to put it in Stefano Rodotà's words. It does not mean that our organs can be replaced or repaired, it means that we may end up being monitored by autonomous processes, thus being *integrally reconstructed* by means of digital machines and their programs.

Whence the need to tackle two sorts of difficulties.

One is that we are not anymore the architects of our own bodily architecture. Perhaps, in a sense, we have never been such architects, since what men actually are has always been a consequence of the necessities and affordances of social life and their coping with them. But there has always been the possibility to understand our alienation and its own logic. Such an understanding of the processes at stake in the formation of the self is not possible anymore, if we take for granted that machines integrated to our body or extending its powers indefinitely calculate discrete quantums of our existence and automatically issue continuously updated profiles of our 'truest' identity.[9] Paradoxically, a 'cyborg' body is even closer to our personality and identity than a natural one, since it allows for an algorithmically, exactly drawn, and constantly updated from of identification.

The *second* difficulty is that an 'electronic body' is not one without history, but one with multiple interrelated and interleaved temporalities. Saving the different states of a 'cyborg's' actions does not simply encompass their succession, it also entails the *superimposing* of any past state of the individual over any other state of the same. A chip does not have the same kind of memory as a body and a brain. A body's memory is its history and individual evolution or ageing. A chip's history is one of saved and computed data, that is a history of endless arbitrary stockpiling. While the natural body replaces itself in a logic of intrinsic biological otherness, the 'cyborg' body gives way to an indefinite growth of data potentially preserving *all* the various states of an individual. Thus in the latter case, otherness means *accumulation* rather than *differentiation*.

'Coupled with electronic devices', Stefano Rodotà writes, 'the body becomes a new object'. Understandably so. Yet this should not be interpreted only in terms of a technological augmentation of our lives and possibilities, as when a disabled man is re-enabled with the use of specially crafted legs.[10] The newly risen questions concern the radical integrity of our body in its relation to machines. Traditionally, we've always thought of machines as mechanical resources to enhance our grasp of things in general and others in particular (exploitation of Nature and Man). Newer technologies have increased the power of machines. How? By allowing them to *monitor* the individual into whom they have been implanted and to whom they are consequently connected. The problem is not that we have to rely on machines to do certain things – indeed some sort of winged or air-sustained machine is required to fly from London to Brussels! The problem lies within the miniaturisation and dissemination that might allow machines to take control over many a decision and action of ours as well as over the way we envision and build our own world. The *relation* between man and machine becomes flawed in the sense that there is actually less and less of such a 'between', as the machine slowly becomes – or is bound to become – an intrinsic part of the body. Being a 'cyborg' has probably very little to do with Hollywood fantasies such as those previously evoked. But it has very much to do with the trivialisation of physically added computer technologies and the non-critical use of data collectors and wireless connectors.

Personality constructing: legal constraints

Adding technology to the body goes further than simply augmenting its power to lift heavy weights or move and run faster. 'Miniaturisation and dissemination' imply that the machine is not visible anymore and that in a way it stops being what we are used to calling a 'machine' – a complex mechanical set of pieces materially attached to one another in order to produce that for which it was built. If computers may still resemble this, the pieces of 'machinery' a body may carry within itself and be connected to are of another nature. Their dissemination relies on both their miniaturisation and their interconnectedness, thus on the digital swarm they end up forming thanks to the reliability of their implemented programs. Which is not a set of empirical properties alone. 'Miniaturisation' means that machines may be rather imperceptible thus consciously untraceable or at least unnoticeable. Things being as they are, it looks as if we were carrying with us not one or a few digital artefacts but a system of machines made out of machines, multiple connected connectors forming systems of functions carried out far above – or under? – what we may individually imagine in terms of data collection and personal profiling.

As a consequence, we reach at this point two series of problems.

a) Nowadays machines help tag and network an individual in many different ways, leading to an expansion and dissemination of what we may determine to be his or her 'private self': his or her personal and social co-ordinates, his or her beliefs, tastes, ideals, plans, etc. If we were to describe the reality beneath such a digital checkering of our so-called 'self', it would have to be depicted in terms of partial signifiers algorithmically crunched and precariously constructed and recon-structed by automated processes all through 'cyberspace' and its various 'places', institutional, public, corporate, private. As they are constantly being tagged, persons are networked, not necessarily between themselves but rather obviously to interconnected servers passing on private information from one point of the network to the other. Resulting from algorithmic treatment, our 'personality' can be treated as negotiable data. Private as it may be, our 'personality' is in this way *publicised* though not exactly made public. Here, the paradox lies in the fact that the circulation of data describing us is a form of publication without publicity. The value of such data consequently comes from their potential though not immediate availability. 'Who' we are, an algorithmically defined person, *must* be computed to represent some transactional value. The private 'us' – our network moves for instance and their digital trails, the will and desires or intentions they describe – is transformed by digital processes to be transmitted to interested parties, commercial or institutional. Extracted from our private sphere and utilised, all that characterises us is re-privatised, though neither with our attention nor to our intention. Circulating among various private and public networks, images or conceptions of our 'selves' are publicised, in the sense they are agglomerated into exchangeable digital goods; and at the same time they are non-published and re-privatised as corporate and sometimes State 'properties' or 'belongings'. This is what a 'person' is in our digital age: a set of patented or registered algorithmic instructions, thus constructions, usable, negotiable, expendable.

Within the context of a post-modern, computed world, the 'person', then, has to be redefined as a *mix* in which 'quality' or 'essence' takes no part. The formal, even metaphysical, background onto which 'personality' was projected in classical times (Hobbes 1981: I, 16) has vanished to the benefit of functional, not even procedural, processing and virtual manufacturing of individual characteristics. Our 'personality' is what algorithms make of it: anticipated responses to commercial proposals or administrative requests.

Indeed, *on the one hand*, 'personal identity is expanding', to write it in Stefano Rodotà's words. This means that we are losing our privileges as 'owners' of our 'self' and are no longer the centre of gravity of any discourse contributing to 'define' who we are psychologically or what we are socially. It has been for a very long time reasonable enough to assume that our 'personality' was characterised by a series of depictions, short or detailed, united into our own consciousness and the knowledge of others corroborating it (Leibniz 1996: II, 27, Hume 2000: I, IV, VI). There was no question there might have been nothing more to the 'definition' of our personality: a set of temperamental properties describing us as wholes, always coinciding within the restrictive volume of a body and the relatively narrow space of an individual's social interactions. We might have had to add the formal requirements of the Law and the more objectionable though ethically useful categories of any given moral theory: 'dignity', 'responsibility', 'subjectivity'. Whatever the compendium, our personality has for long been narrowed to shared discourse of varied sorts from philosophical elaborations to day-to-day beliefs.

Now what Stefano Rodotá characterises as the *expansion* of our 'personality' seems to follow the dynamics and digital dissemination of discourse meant to describe and define us. The use of networks and the wilful participation in multiple social spaces create data sediments into which people and more easily machines may dig to gather whatever information is needed for any purpose of theirs. While our 'personality' used to be discourse *proper*, its reticular expansion has made it mere *information*: packets of objective data that machines may compute, calculate and reassemble – thus redefine – according to their own requirements. As a consequence, in the context of contemporaneous computer practices, our 'personality' should be considered a 'mix' not only in the sense that others contribute to the describing of who we are – this has *always* been the case – but in the sense that automated processes *fabricate* commercially or socially or institutionally useful shapes of our 'selves' and crunch them into machines very often able to determine what we actually choose to think or do – or supposedly so.

But then, *on the other hand*, the question remains and should be asked: what exactly is the informational *matter* that is disseminating and expanding, and that we seem to be compelled to call our 'personality'? It is true that machines replicate all over the network such actions as 'clicks', 'searches' or whatever 'browsing' we have been undertaking.[11] It is also true that tracing online activities helps profiling individuals thus describing the contours of their 'personality'. Such evidence notwithstanding, it is also true that the 'spreading' brought up in the argument pertains to our presence on the networks and the way such presence is *written*. On the networks, our 'personality' draws as much on our behaviour as on its objective *scriptural* form of existence – all the actions we are taking as users of the Internet leaving traces in the shape of 0s and 1s *written* on hard drives and all sorts of digital devices. 'Clicking' is writing as much as it is acting. Our online presence is deployed in the material form of innumerable 0s and 1s first inscribed then moving from one hard disk to another. Which is not at all the same as 'being' in the eyes of others and their own discourse. The contingencies of vision and actual speech have different properties from the sequentiality and order of numeric data. From consciousness to consciousness our personality is blurred, while it is transcribed or algorithmically altered from hardware piece to hardware piece. With expansion comes precision – but also our own *re-writability*. Networks make us exist syntactically while we long for existing semantically. The switch from the former to the latter requires us to re-appropriate the networks. Freely or at least wilfully modifying the available data is difficult and should be helped by proper regulations.[12] Yet it is not only a matter of regulation but also one of education or literacy. A proper consciousness of the stakes and a proper will to be one's own 'personality writer' are both required to shift the present's trend from usability of the person to its full online presence and benefit.

b) A second issue we need to face involves an apparent discrepancy between the perspective of the Law as formulated by Stefano Rodotà when he writes that we need 'to set the conditions for defining [one's personal] identity',[13] and the philosophical approach to the problem of our 'being' as 'persons'. Unless we were to believe in the *Cogito* and its metaphysical meaning and implications, we need to face the fact that our 'personality' carries no substantial properties whatsoever. This does not mean that 'being a person' is an absurdity; it means that 'personality' is a discursive construction rather than a formal expression of qualitative or existential properties. Whence the discrepancy between Law and Philosophy. A philosophical approach to how we face the world and others may have to insist upon the very complex interconnections between various forms of discourse, private and public, legal or fictional, collective or individual. On the contrary, a legal approach to the same problem *must* insist upon the irreducible substantiality of the 'person' as a responsible and free actor in society and among others.

Paradoxically enough, our participation in the life of the networks is just a newer and technologically determined iteration of what might have been meditated in pre-technological terms of sharing values, socially constructing images of the Individual, legends, myths, and the like. It is not really what we actually are that counts, it is the name or names we are given by people (Machiavelli 2009: chapters 15–19). Which is precisely what happens on the networks. In cyberspace we are, so to say, called names. And there is nothing unpleasant or unacceptable about that; as consumers, as music listeners, as video watchers or art appraisers, we are formally constructed, described, then targeted as potential buyers or users of digital or material goods. We may disagree with the automaticity and systematicity of the processes leading to the construction of our 'personality'; it remains that what happens is, at a mechanical and automated level, the exact same procedure that has always happened with the way we are socially 'defined'. But there *is* a difference. It is no longer people who say who or what we are, it is machines. But then again, maybe they are doing a better job than people. . .! Peter might 'know' that his friend Paul likes rock music; he might even offer him the latest CD by his favourite band. But Genius knows it even better and could in the blink of an eye build a few compilations of his most beloved tunes to meet his present mood.[14]

In the past two or three centuries, 'personality' has been defined ethically on the basis of the Subject and his or her 'dignity' or 'humanity'. Whence our social and legal responsibility, topped by our moral 'inner self'. Men used to describe persons while judges, also men, used to praise or condemn their actions. But our actions are not mere actions anymore, they are algorithmic operations – like when we share music or software on P2P networks – and it is not people, it is machines that 'say' who and what we are and 'judge' our actions by allowing or disallowing our connections to take place. What then is Law to protect? The efficiency and neutrality of the network, or the privacy and properties of the connected person? In other words: should it still be the 'person' that should be the centre of gravity of social and moral judgements, or should it be the network itself, namely machines and programs?

The fact that machines know us better than friends is the very reason why Law, not Philosophy, should prevail in the context of our digital lives and real social interactions. Concerning what or who we are, machines have *in a way* made truth functionally irrelevant. For there is nothing essentially or even existentially 'true' about us, only algorithmically crystallised multiple personalities that may have a temporary relevance and usefulness. This might be why, practically speaking, we *must* come to the point where 'personality' becomes undisputable. And it is up to legal discourse to reach that point of understanding and fixing the properties our 'personality' should practically bear. Given the discrepancy between legal and philosophical semantics, it is understandable that law makers should be inspired by the openness of philosophical discourse to avoid narrowing down what 'personality' should mean in the age of automated processes and the computed world.

Identity picking: narrative delusions

Body building and its scarification or tattooing occur within the perimeter of its possibilities, the pleasure, pain and will of a person in flesh and blood. The construction of our personality goes far beyond that. Not only does it extend to the limits of ideologies and philosophical or religious convictions, in their day to day or technically administrative or legal expressions, but it has also developed into digital swarms and data transfers mostly invisible to us and our conscious horizon. A 'person' is an artefact, all the more that it is also made of digital artefacts and their syntactical interconnectedness. This implies that there may be no more room for a traditional focus on the socio-cultural defining of our 'identity' – through the eyes of our family, the evaluations of our schools, the feelings of our friends, the effectiveness of business relations and the demands of the State. As a matter of fact, our 'identity' is nowadays redefined not only in formal judicial terms, not only within a grossly descriptive individual or social perspective, but in the context of an ever mutating environment of interconnected computers.

This underlines a tremendous mutation in the narratives entailed by the urge of a technologically updated 'know thyself'. Indeed it is not only that machines are better than people at 'saying' who we are. We have not merely exported or outsourced the conditions under which we try to be what we tend to be. Sticking to one's 'self' has always been a moral challenge, and ideas have always helped at being a *Mensch* or an *honnête homme*. Generally speaking, the problem of identity narratives nowadays lies within the very processing of data concerning us and the multiple discourse that result from that. In other words, the 'databasisation'[15] of identity narratives has outgrown the very possibility for us to actually know what or who we are, since being what we are supposed to be is a matter of calculations and data transfers rather than facing one's 'self' and judging one's thoughts and actions. Machines and programs depropriate us from the strings of our discourse. They don't make us strangers to ourselves, they more obviously take in charge the narratives expressing our self-understanding, thus our 'true identity'.

As a consequence, we should from now on dig more deeply into the problem of the exporting and outsourcing of any conditions leading to the description of our 'true identity'. As Stefano Rodotà puts it, RFID chips and the traceability of any person's existence tend to radically transform our notion of 'identity' into multiple processes of communication, thus making us lose sight of what we have been used to calling 'the subject'. 'Identity becomes communication' he writes,[16] inferring later on that we are experiencing 'a veritable transfer of identity'.[17] This should not only imply that we as individuals have no ability to define and understand ourselves and try to be what we appear to be. 'Depropriation' is not 'dispossession'. The latter means we would have no means whatsoever to contribute to the narratives of our own self; the former underlines the equivocality of contemporary identification processes and the continuity between self-awareness and personality assessment on the one hand, and programming, databases as well as automated processes on the other. 'Depropriation' means *both* the ability we have to construct ourselves *and* a profound ignorance of the intricacy making automated digital processes what they are – and what *we* are.

So identity narratives seem to translate into another two main issues.

a) The first issue has to do with a gap between self-awareness and ignorance, and its significance. Our computed world is one of multiple technical skills and exactness, not cognitive failure. It is at the same time a world of information over-flow and betrays a consistent inability to envision and embrace the consequences of the digitalisation of our lives. In other words, we have never been so well equipped to understand who we are and what we do, since we possess for instance very accurate statistical instruments allowing for the optimal shaping of our intellectual as well as practical environment; yet so inappropriately capitalising on the tools of our post-modernity, these seem to operate without operators and to deploy the discourses about our 'being' without any narrators formalising their depiction of it. Machines, not the people behind them, build the stories – com-mercial or otherwise – shaping our personal as well as social lives. So it seems as if information and communication technologies had triggered a displacement, misplacement, or even loss in the clear and rational idea and narratives of what and who we are – displacement, misplacement, even loss of *subjectivity*.

By *displacement* we should understand that in a computed world the subject, together with the symbolic space of his or her identity, cannot be pinpointed and described *as such*. If machines, not their operators or people managing their affordances, carry out the task of profiling individuals, subjectivity lies within their processing power and their programs rather than human experience and a reason-able approach and interpretation of people's thoughts and actions. A 'subject' is more and more the result of calculations allowing for the description and valuation of his or her network activity. True, who we are is not reducible to our online presence. But it is also true that who we are cannot anymore be independent from the discourse machines compose by calculating and rearranging the traces we leave behind when surfing the networks or completing simple operations such as withdrawing money from an automatic teller machine or riding the metro with a weekly or monthly electronic pass. Subjectivity and the ability to express its characteristics do not anymore lie within the factual space of speech, human rela-tions, even written documents one has to carry to prove one's identity. Language has always been the core medium for identifying and describing people. Not anymore so. Programs do that and automatic processes pour out statistical results that corporations, institutions and States may use for their own legitimate or not purposes. The problem is only partly that the use made of such information should be submitted to legal approval. From a philosophical standpoint, the problem is more clearly that the centre of gravity of our most intimate 'self' does not lie anymore in our experience of speech and our actual communication with others – in fact the signifying horizon of our own consciousness – but in computer-assisted and 'databasised' profiles. In the future, *The Family Idiot* will not be a work of art or a philosophical analysis, it will be crunched out of algorithmically relevant data to satisfy the purposes that algorithms and programs were conceived for: circum-scribing family perimeters, discriminating conducts and sending out appropriate individually tailored information.

Displacement may as a consequence also mean *misplacement*. Where exactly are the 'self' and its descriptive identity? Not in us only, though of course we still have our word to say – or write and blog and publish. Now turning to machines cannot make up for a clear and straight answer. Machines don't act on their own, they are manipulated by individuals and public or private groups. In other words, their programs technically translate cultural teleologies, either political or economical or social. The *raison d'être* of our most intimate 'self' appears to be hidden in the core of such teleologies. It is indeed extremely difficult to figure out how people are profiled and according to what principles or rules if not merely 'political' or 'commercial'. Once it is said that a given company or State has the technological means to create and exploit individual profiles, nothing much is said about the actual teleological as well as ideological contents of such implemented rules. Admitting that Google or Yahoo or Microsoft do calibrate their users and make profit out of it is not enough to clarify the precise logical trends of such a calibration. In other words, efficiency does not mean understanding, and the technical design of who or what we are does not imply that the process from which it results is understandable. Misplacing our 'self' means that we don't really know where to look to understand the way or ways our identity is sketched then detailed.

Whence its *loss* and our own relative strangeness to ourselves. Being defined and instantaneously redefined according to the necessities of an algorithmic matrix makes us relatively blind to how we act, behave and evolve in a computerised as well as networked environment. Computer skills are varied, but they alone do not make our experience of the networks meaningful. A main reason is that our experience of the networks is above all an *experience* and much more complex than the few 'clicks' or other operations needed to make use of a browser or any other appropriate program. *Being online* means being exposed to a great variety of signifying objects that describe not only an outside reality but also the ways we apprehend it and make cultural projections through it. New semantic and cultural structures have grown out of the networking of data, opinions, images, etc.: traditional sources of knowledge, evaluation, validation, have been outnumbered by multiplied and competing discursive frames – web pages or blogs being the most trivial of them – in such a way that it has become extremely difficult – maybe structurally impossible – to mirror oneself and one's ideas into a fixed cultural pattern. Our loss of identity does not concern us in a personal way only. It rather expresses the instability and dizziness we are bound to evolve from now on due to the digital dispersion of many cultural and ideological beacons.

b) The second issue related to identity narratives concerns our freedom and new ways to loosen many alienating ties to a computerised society and its implied surveillance practices. Paradoxically enough, while we may be quite unaware of our own being online and its 'truthfulness' or 'authenticity', machines do profile us as users of various services, not only commercial but also social and/or political. Our own ignorance is doubled by dispersed though actually hard-memorised knowledge – all the traces we disseminate and leave on computers scattered around the world. Our being is thus memorised as well as kept away from us, as if we were actually not *here* but *there*, anywhere or everywhere: bits of information passing through one server to another at the request of unidentified automatic processes. Which in a way may sound 'big brotherish' enough, but also outlines an interesting aspect of our being, not online only but off-line as well: we are passage, flow, event-fulness, lack of substance. Oddly enough, while our own depropriation may seem frightening – machines mediating more and more consistently our characteristics and identity – the computer grids into which we are being inscribed also stand as a metonymy for our intimate reality and shows us our true 'self' as a mutable puzzle coordinated by external and only partly understandable norms and forces.

In return, this situation gives us a rather stimulating sense of our computer-permeated *freedom*. Freedom has to do with norms as much as with our will. From Hobbes to Rousseau and Locke, from Spinoza to Kant and Hegel – not to mention the Stoics – freedom of thought and action has always been adjusted to the authoritative and legitimate production of the law. Only laws are not anymore just laws. More and more, as Lawrence Lessig has clearly shown, laws are parents to technical rules, formal programming options, scientific problem resolutions, in one word: *code* (Lessig 1999, 2006). Our use of the networks is not merely submitted to the social, ethical, judicial laws under which we are actually living; it is also entailed by the constraints inherent to the algorithms perpetually calculating the paths dispatching our public and private practices through our communication devices. What we do is what code 'allows' us to do, not in the sense of authorised actions but in that of communications made technically possible. Which is not at all the same as boats 'allowing' us to cross oceans or planes to bring continents closer together. What code is 'allowing' has very little to do with the laws of physics and the necessities of nature, and very much so with the ideologies and teleologies of dominant culture, the industrial world and its ideals, a loose and diffuse sense of what good and evil are, etc. Our 'freedom' thus develops within the threads of the networks and their codal intricacy.

A major key, then, to 'freedom', is the way we give ourselves the means to re-appropriate the code and its 'laws': the key to 'freedom' is *knowledge* in renewed forms of digital literacy, in varied forms of computer skill, algorithmic savviness, rhetoric and aesthetic prolixity, etc. It may be ingenuous to say that 'code is at hand' and malleable, but it is also in return authoritative and ideologically objectionable to pretend that it draws on an intangible formal necessity. More and more, code is making us; then more and more we should become aware of the fact that it is akin not to natural necessity but to our teleologies and will. Considering as unquestionable that culture is permeating the 'self', we need to consider and become aware of the fact that from now on our culture is going to be the code we'll be able to invent to re-invent ourselves.

Conclusions expected?

In the new context of information and communication technologies, 'the BPI nexus' calls for a renewed reflection upon our interpretation of the 'self' and the freedom we care to attach to it. Beyond who and what we are intimately and socially, a more fundamental question should be asked about our actual 'freedom' and how we should sense it and live by it. Freedom at the age of universal computing may be neither in networking, nor in refusing the computerisation of our world, but in 'notworking', to put it in Geert Lovink's words (2005). 'Notworking' encompasses the ability to connect and disconnect knowingly, not due to appealing popular aesthetics, blogging fashions, or shortages in power, hardware failures and socio-political constraints. It represents the *digerati*'s freedom, so to say. Which implies literacy and most certainly cultural and intellectual solidarity, common sense, self as well as social and political awareness – renewed forms of *subjectivity*, *citizenship*, and *consciousness*.

The 'BPI nexus' describes not the proximity but rather the interleaving of Man and Machine. Thus, centring our meditation on the 'BPI nexus' is a way of approaching the problem of our very *being* and its post-modern *formatting*. Our 'self' – whatever we wish to name as such – growing increasingly networked, we need to understand the implications of our connectedness as well as the invasion of our being by technological instruments and their components, not only material but also logical. The problem does not simply lie in the presence within the body of chips and other radiofrequency-enabled apparatus; it has to do with their entailed programmatic capabilities and the various forms of exploitation they imply. The scheme of things may be rather easy to perceive – Man as 'hybrid' – but their fundamentals are not so easily elucidated. Both the Modern and Post-modern times have had in common to offer a reflection upon the problematic improvement of 'the Hybrid', from Mary Shelley's *Frankenstein* (1818) to Villiers de l'Isle Adam's *L'Ève future* (1886) and a century later William Gibson's *Neuromancer* (1984). Echoing them, one might ask: *Do Androids Dream of Electric Sheep?* Everybody knows how Philip K. Dick and Ridley Scott have already answered the question.[18] But what did they actually *mean*, really?

Notes

1   See Mathias (2007) for an approach to the problem implied by the confusion between information and knowledge.
2   It suffices to mention Socrates' portrayal in Plato's *Banquet*, 215a *sq.*
3   Diderot, *Le Rêve de d'Alembert*, Paris, 1769, available online at URL: http://classiques. uqac.ca/classiques/Diderot_denis/d_Alembert/d_alembert_2_reve/reve_d_alembert.ht ml (last visit on 1 May 2010).
4   The same question is asked in more academic words by Rodotà in this volume, pp. 180, 189.
5   This important issue is even more obvious in Paul Verhoeven's *Robocop* (1987), a movie starring Peter Weller. The general meaning of the movie is the same as that in the TV serial, but it is made clearer through the computerised mechanical policeman's meditations and his 'metaphysical' reflections on Good and Evil.
6   It is common psychoanalytical knowledge that heavy forms of neurosis may cause 'illnesses'. A marginal account of this trend is shown in Groddeck (1976).
7   Whence, for instance, Descartes' discussion of the proximity and even intimacy of the body and the soul in his *Meditations On First Philosophy*, 'Sixth meditation' (*passim*) and his analysis of passion in *The Passions Of The Soul*, articles 27 *sq.*
8   See Hegel's 'master/slave dialectic' in Hegel (1977: 196 et seq.).
9   For that matter, see Rodotà's mention of an 'electronic leash' on which people (thieves, rapists, etc.) might be more or less arbitrarily kept, this volume, p. 185.

10   As in the Oskar Pistorius issue cited by Rodotà, this volume, p. 180.
11   Keeping the memory of our online actions is not just a fact, it is also a *legal requirement* for Internet Service Providers even in some democratic countries like France.
12   This is the purpose of the 1978 French '*Loi Informatique et Libertés*' – available online at: www.legifrance.gouv.fr/affichTexte.do?cidTexte=LEGITEXT000006068624&date Texte=20090814 (last visit on 1 May 2010).
13   Rodotà in this volume, p. 183.
14   As of 2009, 'Genius' is a function part of Apple's software iTunes.
15   Pr. Hisashi Muroi (2003) may be considered among the first to have theorised the concept of 'databasisation' in the context of literature.
16   Rodotà in this volume, p. 189.
17   Rodotà in this volume, p. 190.
18   For those readers who actually don't, director Ridley Scott's cult movie *Blade Runner* starring Harrison Ford was produced in 1982 after Philip K. Dick's aforementioned short story published in 1968.

Bibliography

Aristotle (1981) *The Politics* (4th century BC), London: Penguin Classics.
DeMello, M. (2000) *The Bodies of Inscription*, Durham: Duke University Press.
De Mul, J. (2009) 'Moral machines: ICTs as mediators of human agency' published as 'Des machines morales', *Cités*, 39, Paris: Presses Universitaires de France.
Descartes, R. (1998) *Meditations on First Philosophy* (1641), London: Penguin Classics.
—— (1989) *The Passions of the Soul* (1649), Indianapolis: Hackett Publishing Company.
Dick, P.K. (2007) *Four Novels of the 1960's*, New York: Penguin Putnam.
Diderot, D. (1976) *D'Alembert's Dream* (1769), London: Penguin Classics.
Elias, N. (1969) *The Civilizing Process*, vol. I. *The History of Manners*, Oxford: Blackwell.
Groddeck, G. (1976) *The Book of the It* (1923), New York: International Universities Press.
Hegel, G.W.F. (1977) *The Phenomenology of Spirit* (1807), Oxford: Oxford University Press.
Hobbes, T. (1981) *Leviathan* (1651), London: Penguin Classics.
Hume, D. (2000) *A Treatise of Human Nature* (1739), Oxford: Oxford University Press.
Leibniz, G.W. (1996) *New Essays on Human Understanding* (1705), Cambridge: Cambridge University Press.
Lessig, L. (1999 1st ed.), (2006 2nd ed.) *Code and Other Laws of Cyberspace*, New York: Basic Books.
Lovink, G. (2005) *The Principle of Notworking*, Amsterdam: Amsterdam University Press.
Machiavelli, N. (2009) *The Prince* (1513), London: Penguin Classics.
Mathias, P. (2007) 'Power Matters', paper presented on 27 April 2007 at the Centre Marc Bloch in Berlin, available at http://diktyologie.homo-numericus.net/spip.php?article70 (last visit on 21 June 2010).
Mauss, M. (2006) *Techniques, Technology and Civilisation* (translation of idem (1936) 'Les Techniques du corps', (32) *Journal de Psychologie*, 3–4, French text available online at http://classiques.uqac.ca/classiques/mauss_marcel/socio_et_anthropo/6_Techniques_cor ps/ Techniques_corps.html (last visit on 9 November 2010). Oxford, New York: Berghahn Books.
Muroi, H. (2003) 'Problems on Interpretation in the Age of Database', paper presented at the International Congress of Association of Japanese Literary Studies (UCLA, Los

Angels November 2003), available at: www.bekkoame.ne.jp/~hmuroi/e11.html (last visit on 1 May 2010).

Plato (2008) *Banquet* (4th century BC), Cambridge: Cambridge University Press.

Renaut, L. (2004) *Marquage corporel et signation religieuse dans l'Antiquité* (Thèse de doctorat), Paris: École Pratique des Hautes Études.

Epilogue
Technological mediation, and human agency as recalcitrance

Antoinette Rouvroy

Autonomic computing is nothing but a projection, or an evolutionary program for the 'computational species' – a species composed of technologically mediated subjects (Verbeek), increasingly intersecting and interacting with digital systems (Rodotà). Yet, its foreshadows plunge us into abyssal ontological, epistemic and normative interrogations with regard to the fate – and thus also the actuality – of human identity, agency, autonomy and legal subjectivity. As such, autonomic computing, as a vision, revitalizes ancient questions. It reminds us of a series of uncertainties regarding the implications, for the law, of conceiving human identity and singularity as a dynamic process rather than as a fixed phenomenon; it reiterates the question whether a subject can be said to be autonomous and responsible for his or her actions despite not having a fixed identity, and despite not being in control of the conditions and circumstances which shape and re-shape his or her autonomy (Rodotà); it evokes the doubtful connection or articulation between the concepts of body, person and identity (Mathias); it renders the question crucial of what residual or fundamental role embodiment does, could or should play in a context of digitalization of life itself, or whether embodiment still – or at last – attests to the 'nature' of human agency (Hyo Yoon Kang). These interrogations recall the precarious nature of human intentionality, reasons, motivations and actions and reinvigorate questions about the role – either real or fictional – the law presupposes these 'capabilities' to endorse as constitutive of or conductive to human agency.

It is not my ambition, in this epilogue, to try and summarize the highly sophisticated, nuanced and thoughtful contributions gathered in the previous chapters. Expecting the flamboyance of an afterimage, of a structuring 'motive' for the contrapuntal articulation of the philosophical registers visited or re-visited in this volume would of course be naïve and premature given both the launching and prospective character of the 'subject'. One safe bet one can make though is that such interdisciplinary encounters, inaugural as they appear, will be paramount as to preserve, collectively as well as individually, the possibility to reflect about and evaluate the coming transformations of our relations to the world and to ourselves. Preventing the anticipatory erosion of philosophical vigilance, philosophical puzzles happen as interruptions, suspensions, or retardations of the otherwise unquestioned – and automatic – shift from anticipation to actualization of technological 'projections', be they called autonomic computing, ambient intelligence, or ubiquitous computing. As such, encounters of the type attested in the volume, and, ideally, their accommodation to a wider audience, should afford polities the space and time necessary to govern technological and social evolutions according to deliberate, explicit and sustainable projects, and to avoid having world visions exclusively embedded in strong, yet often hidden, commercial interests imposed upon themselves. The inspiring idea of a community of rights suggested by Roger Brownsword is particularly relevant to show that making projects, beyond or despite the apparent immanence of the digitalized society, is still peoples' and communities' responsibility. Let us not forget, in this regard, that most of our post-industrial artefactual environment, seamless as it appears and translating our own embodied life into dis-embodied data-sets, have been and are produced by working and suffering bodies elsewhere, in 'third-world' countries where most people, struggling for their physical survival, do not have access to the brave new data-world we are contemplating with both excitement and fears. Among the most urgent issues, which is nevertheless beyond the reach of this volume, and which would require attention and intervention of other disciplines than philosophy of law and philosophy of technology, is the somewhat 'cannibalistic' exploitation of invisible bodies in the so-called 'under-developed' parts of the world, whose agency consists mainly in struggling for their own physical survival, and the contrasting escape from embodiment experienced in our post-industrial informational capitalistic society.

This volume gathers contributions to the study of 'agency' in a world of autonomic computing, with a tacit assumption situating the 'problem', or 'enigma' of the subject (Rodotà) in the context of so-called post-industrial liberal democracies. The reflections provided in this book are obviously impressed by a Western, liberal culture. The questions the vision of autonomic computing suggests to philosophers of technology and to philosophers of law are also raised from within the same culture where the autonomous individual appears both as a presupposition and as a normative project in itself. In such a context, it is indeed highly relevant to ask what autonomic computing would change – if anything – with regard to human intentions, reasons, motivations, embodiment, language, actions and the articulations of all these 'attributes' intuitively associated with 'agency'. It is also highly relevant to ask how such changes would in turn impact on the liberal legal order, which presupposes, and locates at its core the unitary, intentional, rational, conscious, embodied, and speaking individual. And there will never be too many interdisciplinary conversations about these topics.

In the following pages, I would like to suggest a hypothesis about the manner we ask such questions, with methodological and substantive consequences. I would like to try rephrasing the question of human agency as a question about 'recalcitrance', or about 'excess' rather than as a question about 'control' or 'intentions'. From reading the various chapters of this book, I got the very subjective impression that there is something in human beings that machines will never succeed to either anticipate or regulate, something that happens in excess to the 'traces' we leave in registers and databases (Durante), even if these traces are re-assigned to us as a destiny. That impression filled me with the irrepressible joyful sensation that *potentiality* – the intempestivity, spontaneity, unpredictability, that is, all these vital qualities which Hannah Arendt associated with natality – will always withstand *probability*.

This may appear counterintuitive given the current efforts – in terms of funding and ingeniosity – deployed in both the public and the private sector in order to develop smart, intelligent, systems of detection, classification and forward-looking evaluation of human behaviours, attitudes, preferences, propensities, etc. and to smooth human experiences and interactions in unprecedented ways. A recurring theme in the volume depicts human agency as unavoidably mediated by technology (Ihde, Verbeek). Intervening as 'intuitive and unobtrusive' technological cognitive interfaces, these technologies may enhance social legibility, self-reflexivity, the capacity and skill to evaluate alternative courses of action, and thus increase autonomy and accountability (Kallinikos). Yet – and this may be a specificity of the actual and forthcoming ICT interfaces which places them apart from ancient tools and instruments – they do so through humanely unintelligible algorithmic processes. Fundamental epistemic questions arise when humans implicitly give up the ambitions of modern rationality linking observed phenomena to their causes, and privilege an algorithmic – and, in this sense, post-modern – (ir)rationality, rendering the world *insignificant* but *predictable* according to a purely inductive (based on correlations) and highly effective statistical logic.

That the kind of knowledge emanating from algorithmic processes appears to escape traditional knowledge validation tests does not necessarily result in such knowledge being 'unilaterally' imposed on human agents though, and the norms (criteria of normality, desirability, dangerousness, needs. etc.) ensuing from the statistical recording of 'the real' will not unavoidably translate into uncontested or unconsciously implemented normativity. Human agency and human subjectivity oppose a series of 'recalcitrances' to their own previsibility, anticipation and pre-emption by autonomic computing systems and their precursors. Mapping the possible zones of recalcitrances, the places which shall remain untouched by autonomic computing never mind how broad and multimodal the reach of such systems would be, might well give a few indications about what human agency *is*, and of what it *can do*. At a time where efficiency, previsibility and risk minimization have become leitmotivs in most political and industrial agendas, one may also wonder about the function such recalcitrance may have in the project of a community of rights and as part of the normative metabolism of liberal democracies.

That the ingenuity of human behaviours will always, in part, escape both predictability and regulation by technology (Don Ihde) and the fact that the 'double hermeneutics' described by Don Ihde allows individuals to realize how machines profile them, and to adapt their behaviours accordingly (obeying or disobeying the norms of expectations 'of the machine'), leading to a more complex understanding of the epistemic processes at play,[1] leaves the question open of what the individual and collective meaning and value there is (if there is such value) in this escaping ingenuity. The question is not absolutely trivial: should one continue to try and improve technologies that channel human behaviours and decrease the margins where such ingenuity may express, or should one rather privilege other projects, building on the idea that human ingenuity is indeed something that must be encouraged and protected, be it at the cost of absolute previsibility? Responding to such a question is indeed a fundamental precondition to set research agendas in an era of converging sciences and technologies (where neuro-sciences, ICTs, network technologies and the whole range of bio- and nano-technologies might soon be combined as to decrease the rate of 'recalcitrance' and of the associated spontaneity and unpredictability).

Recalcitrance does not necessarily presuppose a notion of control over what 'causes' our behaviours though. The lack of awareness and control over what causes their own actions is nothing new for human beings. Intentions and reasons are never the exclusive nor the ultimate causes of actions. 'The sense of control we have with regard to our own actions is an illusion, produced after the fact', Mireille Hildebrandt recalls. This, however, does not *ipso facto* expel the possibility for human agents to build and project their own *motives*, that is, to ascribe or give significance to their actions *as their own*. I would like to conjecture here that the 'production' – of motives, that is, of meaning and values – after the fact, is what matters for human agency, in spite of – or rather thanks to – the inherent belatedness of this 'production', its non-coincidence with and unfaithfulness to the actions we *motivate* after the facts. This distance between the facts or acts and the words and motivations expressed through the technology of language,[2] this 'inactuality', the unfilled gap between things and words affords human agents their ***potentiality***,[3] which makes them escape determinations or remote controls enacted through technological interfaces (de Mul and van den Berg). This production of significance and value at a distance from the facts and acts is also what allows them to give account of themselves (Judith Butler) either in judicial Courts or in daily lives. The primacy of significance over causality has been advanced, implicitly, by Robert Musil (1956: 613), in his wonderful, unachieved novel written from 1921 until the author's death in 1942, *The Man without Qualities*[4]:

> The motive is what drives me from signification to signification. Something happens, something is said: that increases the meaning of two human lives, that meaning reinforces their union; but what has happened, which physical or legal the event represents, that is unimportant, this is another issue.[5]

Although motives are not causes, they are what gives actions their meaning and value. However – except in a psychoanalytic context where answers are given to questions which have not precedingly been uttered – in daily life, motives are best expressed when the agent is interpellated or addressed by another, or by others.

A series of questions immediately arise such as whether acknowledging the primacy of motivation (as significance, that is, as that which gives meaning and value to an act and as that which, when expressed by an agent giving account of him or herself, contributes to the constitution of his or her identity) over causality would challenge the current privileged position that causality occupies in legal reasoning..Another inescapable question would be what exactly, in a context of autonomic computing, would be the role of language, verbalization and embodiment – another source of recalcitrance and relationality, allowing a consideration of human legal subjects as material bodies with embodied minds rather than as computable minds with absent bodies (Hyo Yoon Kang).

Many things and ideas remain to explore with regard to the *kinds of addresses* directed at individuals in an era of autonomic computing. Let's start from the most improbable scenario where autonomic computing, articulated with intelligent environments, succeeds in regulating human activities so that human conflicts, criminality and disobedience disappear – which, as Don Ihde explains, is mostly improbable – and which would then also dispel the need to interpellate one another, or to oblige or allow agents to give account of themselves in courts. The only 'other' to whom individuals would be called to give account would be the autonomic computing system itself, and this would probably not be through verbal interaction but through pre-conscious mutual attunement. Yet, it is not certain that subjects preexist *qua subjects* to their 'interpellation' by 'others', Althusser, Foucault, Lacan, Duster, Butler and a few others have argued. If it is the case that subjectivity is unavoidably relational (a matter of giving account and being recognized), what happens when the 'other' to which one has to give account of oneself is a system of autonomic computing? This is of course the most implausible scenario, as already mentioned, and its only merit is the merit one sometimes recognizes in caricatures or ideal types.

A second scenario, which is much closer to the current situation, is a scenario where operations of collection, processing and structuration of data for purposes of data-mining and profiling, helping human agents to cope with circumstances of uncertainty or relieving them from the burden of taking decision in routine situations, have become central to public and private sectors' activities. Such systems indeed interpellate human agents through the myriad of dis-embodied, decontextualized data-points or networks of localizations into actuarial tables of various kinds. I have depicted (Rouvroy), in this volume, the difficulty for individual agents to give account of themselves both individually and collectively whenever they are addressed through dispersed profiles constructed according to opaque algorithms ignorant of personal autobiography and of socially experienced communities. The question becomes, then, whether and how the legal order should preserve mechanisms whereby *motivations* can still be heard. The intensification of datamining and profiling, I have argued, brings forth an 'algorithmic normativity' which appears as a « natural » germination from the digital transcription and statistical analysis of 'reality', and therefore resists characterization as either spontaneous or artefactual. As such, the resulting norms elude usual tests both of epistemic validity and of political legitimacy. Yet, when embedded in systems of detection, classification and anticipative evaluation of human behaviours, data-mining and profiling methods indeed have governmental effects in the various spheres where they apply:

bypassing inter-individual interactions between gatekeepers or governants and individual subjects, they ease decision-making processes by dispensing from individualized evaluations of deserts, merits, abilities or needs. But such an 'irrational rationalization' (rendering the world predictable but insignificant), useful and emancipatory in certain circumstances, may become threatening for human agency whenever it deprives individuals from the possibility to give account of themselves to 'others' which are themselves capable of recalcitrance and therefore offering the subject a mirror allowing self-recognition as recalcitrant subjectivity.

What remains certain, anyway, is that the theme of 'autonomic computing and transformations of human agency', recalcitrant as it is itself to any kind of definitive conclusion, provides an ideal scene for the vitalizing confrontation of world versions and visions.

Notes

1   That is also why – from the most trivial to the most complex technological *dispositive* – none of the 'autonomic' systems one may imagine will ever take over human intervention, according to Don Ihde.
2   See note 4 of the introduction by Mireille Hildebrandt.
3   On the notion of *potentiality*, which I believe is fundamental to pursue our inquiry into the impacts of autonomic computing for human agency, see Agamben (1999). See also the distinction made by Pierre Macherey, following Spinoza, between *potentia* (puissance) and *potestas* (pouvoir) in Macherev (2009).
4   Translations of *The Man Without Qualities* in English have been published by Ernst Kaiser and Eithne Wilkins in 1953, 1954 and 1960, and by Knopf in 1995.
5   My translation of 'Le motif, c'est ce qui me conduit de signification en signification. Quelque chose arrive, quelque chose est dit: celà accroît le sens de deux vies humaines, ce sens renforce leur union; mais ce qui se passé, quelle notion physique ou juridique l'événement représente, cela n'a aucune importance, c'est une toute autre affaire.'

References

Agamben, G. (1999) *Potentialities. Collected Essays in Philosophy* (ed. and transl. with an introduction by Daniel Heller-Roazen), Stanford: Stanford University Press.
Macherey, P. (2009) *De Canguilhem à Foucault. La force des normes*, Paris: La fabrique editions.
Musil, R. (1956) *L'Homme sans qualités* (transl. from German by Philippe Jaccottet), vol. II Paris: Seuil, 1956.

Index